The Configuration Space Method for Kinematic Design of Mechanisms

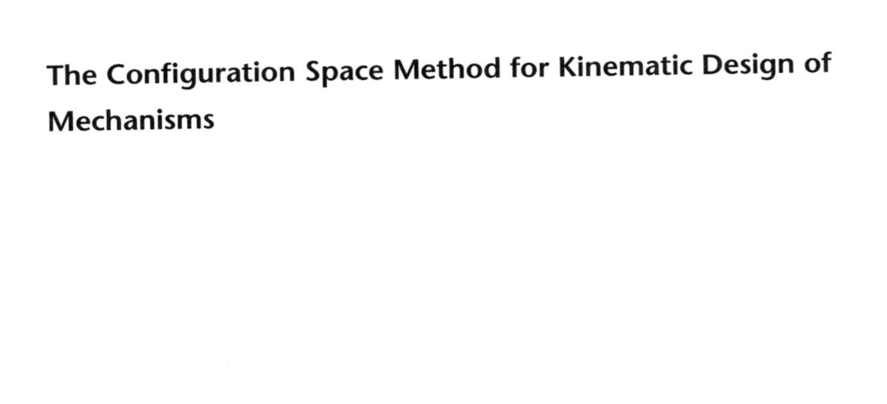

The Configuration Space Method for Kinematic Design of Mechanisms

Elisha Sacks and Leo Joskowicz

The MIT Press
Cambridge, Massachusetts
London, England

For information about special quantity discounts, please email special_sales@mitpress.mit.edu

This book was set in Stone Serif and Stone Sans on 3B2 by Asco Typesetters, Hong Kong. Printed and bound in the United States of America.

Library of Congress Cataloging-in-Publication Data

Sacks, Elisha, 1958–
The configuration space method for kinematic design of mechanisms / Elisha Sacks and Leo Joskowicz.
p. cm.
Includes bibliographical references and index.
ISBN 978-0-262-01389-5 (hardcover : alk. paper) 1. Machinery, Kinematics of. 2. Computer-aided design. 3. Configuration space. 4. Machine design.
I. Joskowicz, Leo, 1961– II. Title.
TJ175.S23 2010
621.8′11—dc22 2009030466

10 9 8 7 6 5 4 3 2 1

I dedicate this book to my wife, Jennifer, the light of my life.
Elisha

I dedicate this book to my daughters, Dana and Yael, to my parents, Esther and Alfredo, and to all my close friends.
Leo

Contents

Preface

This monograph presents the configuration space method for computer-aided design of mechanisms with changing contacts of parts. Configuration space is a complete and compact geometric representation of the motions and interactions of parts in a mechanism that supports the core design tasks of analysis, synthesis, and tolerancing. Our method is the first general algorithmic treatment of the kinematics of higher pairs with changing contacts. It helps designers detect and correct unexpected kinematic behaviors and design flaws, as demonstrated in automotive, micromechanical, and optical case studies.

This book is intended for students, researchers, and engineers in mechanical engineering, computer-aided design, computer science, and robotics. The presentation is self-contained and is suited for a course, a seminar, or independent study. The prerequisites are freshman mathematics and computer science.

The first part describes the configuration space framework and the algorithms for the kinematics of mechanisms. The second part describes the algorithms for kinematic analysis, tolerancing, and synthesis based on configuration spaces. The third part presents four case studies taken from industry in which the configuration space method supports the analysis and design of mechanisms. A catalog of higher-pair mechanisms is given in appendix A. Appendix B describes an open-source C++ mechanical design system, called HIPAIR, that implements some of the configuration space methods described in this book, including visualization of configuration space and kinematic simulation. HIPAIR comes with an interactive graphical user interface and many sample input files for mechanisms.

Acknowledgments

We wish to thank the many people who have contributed to this book. Min-Ho Kyung at Ajou University, Sowon, South Korea, developed the parameter synthesis and optimization algorithms. Ku-Jin Kim at Ajou University developed the partition algorithm for a spatial fixed-axis configuration space. Vijay Srinivasan from the IBM T.J. Watson Research Center, York town Heights, participated in our early research on contact zones and Min-Ho Kyung participated in our research on nonlinear contact zones. Ralf Schultheiss and Uwe Hinze at Ford Motors, Cologne, Germany, introduced us to automotive design, provided the manual gearshift case study, and supported our research with a Ford grant. Jim Allen, Steve Barnes, and many others at Sandia National Laboratories, Albuquerque, New Mexico, introduced us to the design of MEMS mechanisms and provided the case study of the torsional ratcheting actuator. The National Science Foundation in Washington, D.C. and the Israel Science Foundation in Jerusalem supported this work. Ruby Shamir from Hebrew University of Jerusalem provided valuable comments on the first part of this book.

I, Elisha, wish to thank my parents for raising me to love books and independent thinking. I enjoyed and learned from discussions with many people, especially Hal Abelson, Jim Allen, Bruce Donald, Mike Erdman, Pat Hanrahan, Chris Hoffman, Matt Mason, Victor Milenkovic, Tomas Lozano-Perez, Ralf Schultheiss, Herb Simon, Gerry Sussman, and Ken Yip. Leo Joskowicz introduced me to mechanism design and to configuration spaces. He hosted me for many years in Paris, New York, and Jerusalem, and has always been a loyal friend. My wife, Jennifer, and daughters, Sarah, Talia, and Anat, encouraged me along the way and offered marketing advice. Sad to say, we will not be able to use "Passion in the Desert" as the title, nor can a romantic oasis scene appear on the cover. I realize that equations and diagrams are sorry substitutes. Last but not necessarily least, my cat,

Fluffer, has kept me company (often asleep on the couch) for most of the long hours that I spent writing and revising this book.

I, Leo, wish to thank my parents; my daughters, Yael and Dana, and my many beloved friends around the world for their support and patience in good and bad times. I thank Elisha Sacks for being a perseverant and outstanding partner in this long journey. I also want to thank Sanjaya Addanki, Sesh Murthy, Scott Penberthy, Jarek Rossignac, and former colleagues at the IBM T.J. Watson Research Center, Yorktown Heights, NY, for their support and fruitful discussions on mechanical design and research. The Institute of Computer Science, now the School of Engineering and Computer Science at the Hebrew University of Jerusalem, Israel, provided an optimal environment for writing this book.

The Configuration Space Method for Kinematic Design of Mechanisms

1 Introduction

Mechanisms are assemblies of moving parts that perform useful tasks. They are pervasive in modern life and include familiar mechanical systems, such as door locks, transmissions, and gearboxes, and specialized ones, such as industrial robots and microscopic mechanisms fabricated on silicon chips (MEMS). The designing of mechanisms is an important engineering task that motivates research in design methods and in computer-aided design (CAD) tools.

A key aspect of this task is kinematics. This is the branch of mechanics that studies the motion of parts independently of the forces acting on them. It assumes that parts are rigid bodies with fixed shapes. A mechanism performs a task by transforming input motions into output motions through the contact of parts. The transformation of motions is called the kinematic function of the mechanism. The design goal is to ensure that the intended and actual motions match. Studying the kinematics is an early and crucial step in designing a mechanism. It helps answer many questions about the workings of mechanisms and is a prerequisite for further mechanical studies involving dynamics, stress, and deformation.

The main kinematic design tasks are analysis, synthesis, and tolerancing. Analysis derives the kinematic function of a mechanism from a specification of its parts' shapes and motion constraints. For example, a rotating gear wheel is specified by its profile (how many teeth and what shape) and the configuration (position and orientation) of its rotation axis. A pair of gears is analyzed to determine its gear ratio. Synthesis is the inverse task of devising a mechanism that performs a specified function. The starting point can be a prior design or a novel design concept. Analysis and synthesis disregard the imprecision of manufacturing, which causes actual parts to vary from their intended shapes and configurations. The tolerancing design task is to determine the kinematic effect of variation in manufacturing and modify the design or the manufacturing process to ensure correct function.

The foundation for a systematic study of mechanism kinematics was laid by the German engineer Franz Reuleaux (1829–1905). Reuleaux defined mechanisms as collections of basic building blocks, called kinematic pairs, and developed a classification system for mechanisms. A kinematic pair consists of two interacting parts. There are two types of pairs: lower and higher. Lower pairs involve a single, permanent surface contact between two parts, such as a round pin in a matching hole. The parts have simple relative motions, such as rotation around an axis. All other pairs are higher pairs. Lower pairs are idealizations of higher pairs because real parts require a small clearance to function.

The compositional model of mechanisms uses a hierarchical analysis strategy: compute the kinematic functions of the pairs, then combine them to obtain the kinematic function of the mechanism. The kinematic function of a pair is determined by the parts' geometry in the neighborhood of the points of contact. Analysis consists of formulating and solving contact equations. These equations are highly nonlinear and can be solved in closed form only in simple cases. Combining a pair' kinematic functions entails an analysis of the interactions among the pair's contacts. This analysis is difficult, even for mechanisms with few parts, because there are many potential interactions. Synthesis and tolerancing are also performed hierarchically and pose similar challenges.

The mainstream approach to managing the complexity of kinematic design is to identify special cases that restrict the interactions of parts. Kinematic pairs are assumed to be lower pairs or to have fixed, closed-form contact relations. Mechanisms are limited to assemblies of such pairs. Consequently, a mechanism's kinematic function is specified by a fixed set of equations that can be solved with an efficient numerical algorithm. The most common categories are linkages, cams, and gear mechanisms. Linkages consist of lower pairs. Cam mechanisms consist of cams in permanent contact with followers. Gear mechanisms consist of meshed gears. These categories exclude many important mechanisms.

Pairs with multiple contacts are more versatile than fixed-contact pairs because they can perform multiple functions through changes in their contacts. Higher-pair mechanisms are typically cheaper, lighter, and more compact than lower-pair mechanisms. Examples include sewing machines, copiers, cameras, and compact disc players.

In this book we present a general computational theory of kinematic design that covers all types of mechanisms. In this approach, the contacts of

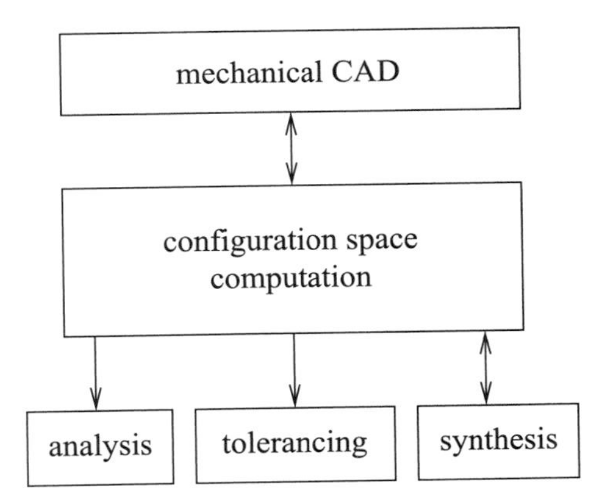

Figure 1.1
Kinematic design of a mechanism using configuration spaces.

parts are studied in a geometric representation called configuration space. We have developed algorithms for analysis, synthesis, and tolerancing within the representation of configuration space. The algorithms provide novel computer-aided design tools that can help designers detect unexpected behaviors, correct design flaws, and study the kinematic effects of variations in manufacturing.

Figure 1.1 illustrates our kinematic design paradigm. Mechanical CAD packages are used to create, modify, and visualize mechanism models. The computation of configuration space generates a kinematic model that is used in analysis and tolerancing, and is modified in synthesis.

The rest of this chapter is organized as follows. In section 1.1, we introduce kinematic function. In section 1.2, we discuss and illustrate kinematic design. In section 1.3, we describe the content and organization of this book.

1.1 Mechanisms and Kinematic Function

The kinematic function of a mechanism is determined by the shapes, configurations, and motions of its parts. For example, consider the indexer mechanism in figure 1.2. It is composed of six parts: a driver, an indexer, a pawl, a pin, a lever, and a frame. The driver is an offcenter cylinder that acts as a cam. The indexer is a gear with 24 teeth shapes as trapezoids. The pawl has a triangular tip shaped to follow the indexer's teeth. The pin is a rectangular block. Each of these parts rotates on a shaft mounted on a cylindrical hole in the frame. The lever has a rounded triangular tip shaped to engage

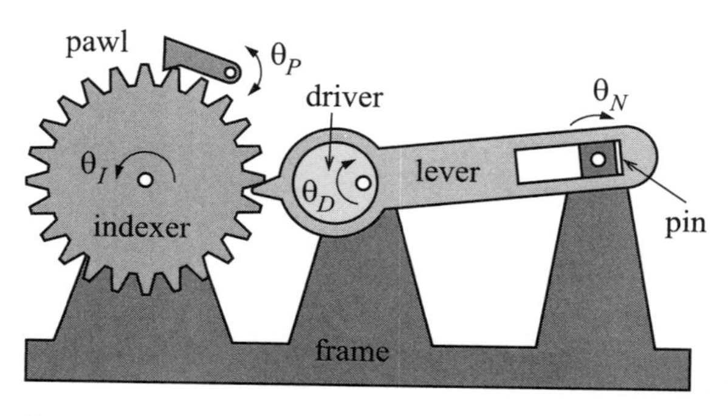

Figure 1.2
Indexer mechanism.

the indexer's teeth, a cylindrical hole that fits over the cam, and a rectangular slot that fits over the pin. Its motion is constrained by the driver and the pin. The frame holds the parts together. The indexer mechanism has six lower pairs (indexer, pawl, driver and pin mounted on the frame, lever-pin, and driver-lever), and two higher pairs (indexer-pawl, indexer-lever). The rotations of the parts are as indicated by the arrows. (The following conventions for arrows are used throughout the book. The intended part rotation and translation directions are indicated with thick-headed arrows. The rotation angle is always measured in the standard counterclockwise direction even when the arrow points clockwise. The translation is always measured along the standard right-handed coordinate axis even when the arrow points left. A double thick-headed arrow indicates back-and-forth motion.)

The kinematic function of the indexer mechanism is to advance the indexer wheel by one tooth (15°) for every turn of the driver. As the driver turns clockwise, the lever tip traces a closed trajectory whose form is determined by the relative position of the driver and pin rotation axes, the driver's offset, and the length of the lever. This causes the indexer to rotate counterclockwise by 15°. The pawl prevents clockwise rotation of the indexer. Figure 1.3 shows four snapshots of one cycle: (a) the start configuration, (b) the lever tooth engaging the wheel tooth, (c) the lever driving the wheel, and (d) the lever tooth disengaging the wheel tooth.

In designing a mechanism, the kinematic function is usually derived first because it sets the stage for the study of other physical phenomena. In our example, dynamical analysis determines the driver and indexer torques based on their mass and shape. Stress analysis determines the indexer tip deformation based on its load and material properties. In both analyses, the kinematic function constrains the motions and contacts of the parts.

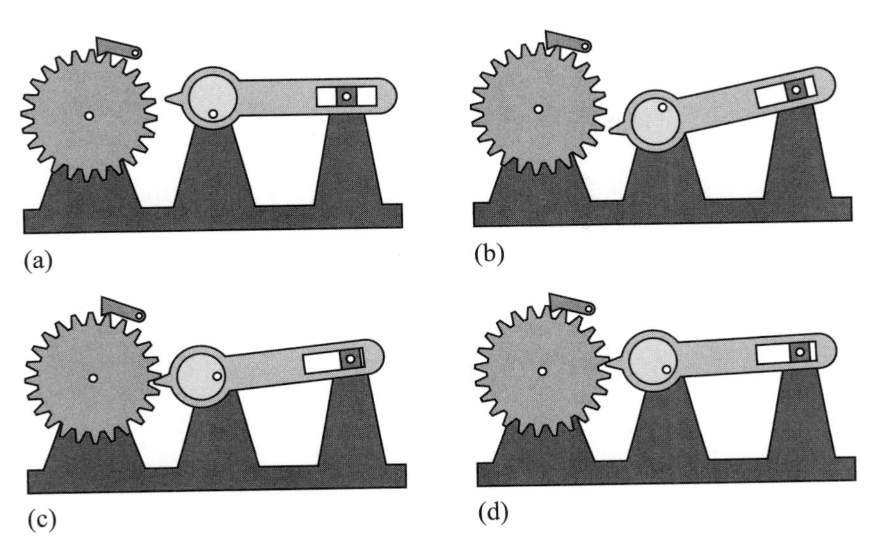

Figure 1.3
Snapshots of the kinematic function of an indexer mechanism.

Kinematic design is an iterative process consisting of five steps (figure 1.4). Conceptual design consists of selecting a design concept that captures the desired kinematic function. The concept determines the structure of the mechanism: the parts, the kinematic pairs, and the intended kinematic function. Parametric design consists of building a parametric model of the mechanism that encodes the shapes and configurations of its parts. Parameter values are chosen to achieve the intended kinematic function in the absence of manufacturing variation. The resulting design is called a nominal mechanism. Analysis, tolerancing, and synthesis are then performed as described earlier. They are iterated until a satisfactory design is obtained. When this is impossible, the parametric model is revised or an alternative design concept is elaborated.

1.2 Kinematic Design

We illustrate the design cycle using the indexer mechanism. Figure 1.5 shows a parametric model for the conceptual design described here. The parameters include the relative positions of the fixed axes, the offset of the driver, and the radii of the indexer teeth. The parts are modeled in the *xy* plane because they have a constant cross-section along the *z* axis. Likewise, their motions are modeled in the *xy* plane. The indexer, pawl, driver, and pin are mounted on cylindrical shafts, so each is modeled with a pin joint.

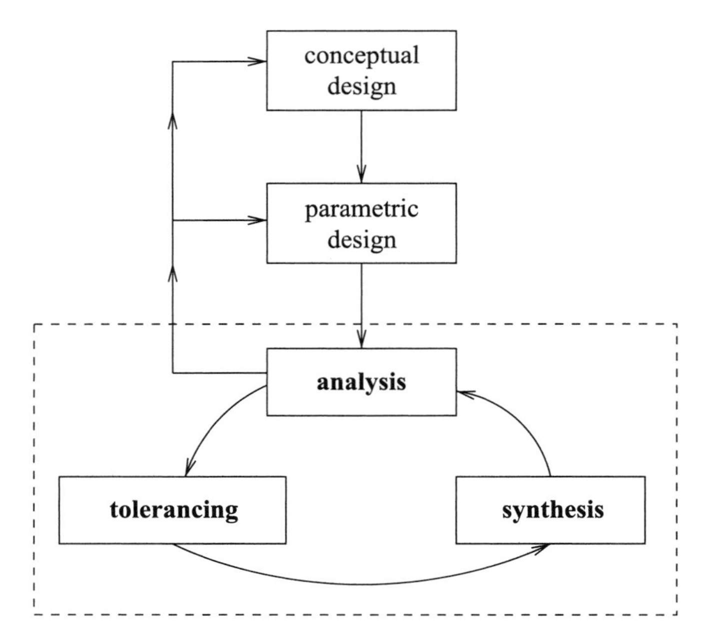

Figure 1.4
Kinematic design cycle. Boldface indicates tasks covered in this book.

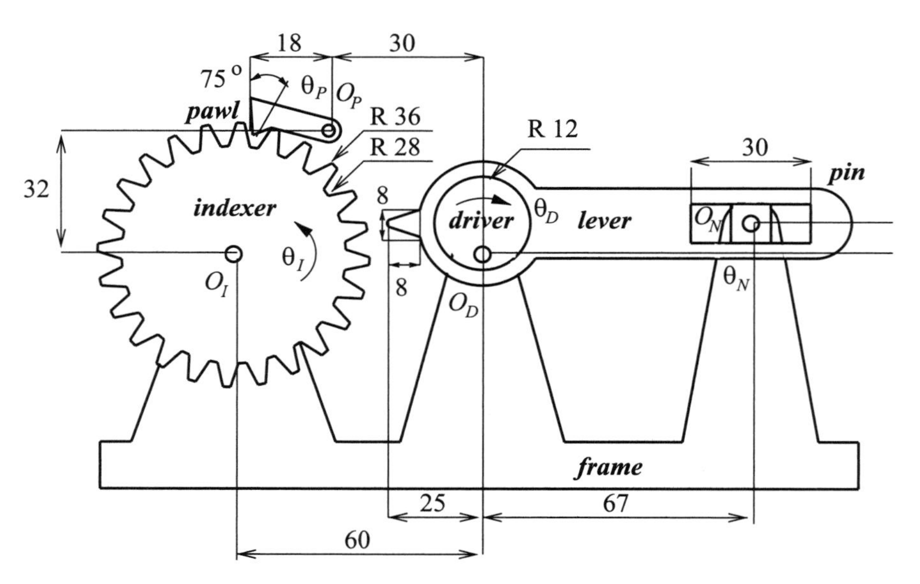

Figure 1.5
Indexer mechanism: functional parameters (in millimeters).

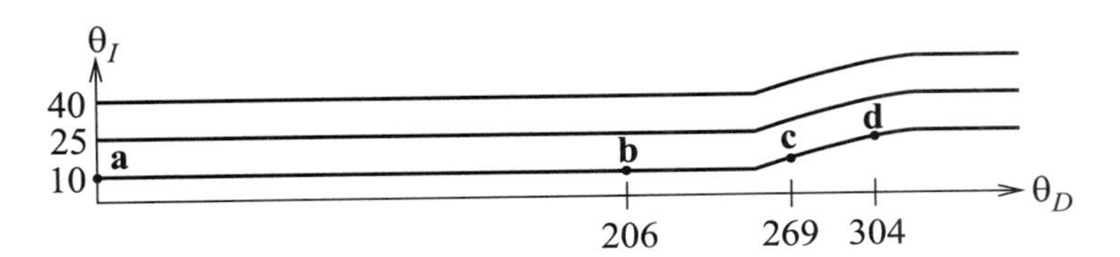

Figure 1.6
Kinematic function of the indexer mechanism.

Their configurations are specified with orientation parameters with respect to fixed axes: θ_I and O_I for the indexer, θ_P and O_P for the pawl, θ_D and O_D for the driver, and θ_N and O_N for the pin. The intended rotation directions are indicated with arrows. The lever is not mounted on a shaft, so its configuration is specified with two position parameters and an orientation parameter. The pin fits into the lever's rectangular slot, so the pin-lever contact is modeled as a slider joint. The driver and lever have a permanent contact between a circle and a matching circular hole. The pawl-indexer and lever-indexer interactions are more complex because they involve multiple changes of contact.

Having created the parametric model, the designer faces several key questions. Will the nominal mechanism function as intended or can it fail, owing to unexpected motions of parts? What modifications are necessary to make the mechanism function properly despite manufacturing variation? What tolerances are required? Can they be loosened, which reduces manufacturing cost, by modifying the nominal design? To answer these questions, the designer analyzes the current design and synthesizes alternative designs.

The analysis derives the kinematic function of the nominal mechanism from its parametric model. The kinematic function is quantified by plotting the relation between the motion parameters. Figure 1.6 shows the plot for our example. The functional relation is between the driver's orientation angle, θ_D, and the indexer's orientation angle, θ_I. The initial values are $\mathbf{a} = (0°, 10°)$. Angle θ_D increases clockwise and θ_I increases counterclockwise. The bottom curve represents the work cycle shown in figure 1.3 with points **a–d** corresponding to the snapshots. The indexer is at rest on the horizontal part of the curve and is driven on the diagonal part. At the end of the cycle, θ_D equals 360° and θ_I has increased to 25°. The middle curve represents the second cycle, in which θ_I increases from 25° to 40°, and the top curve represents the third cycle. Since the indexer has 24 teeth, there are 24 such segments at 15-degree intervals.

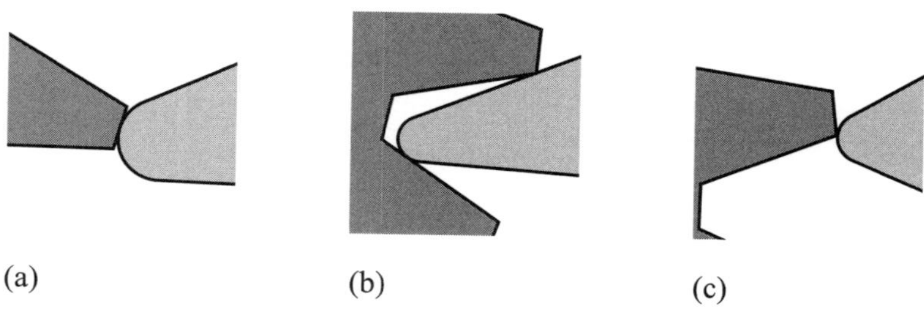

Figure 1.7
Lever pin failures (detail): (a) tip too wide, (b) tip too long, (c) tip too short.

The kinematic function rarely has a closed form, especially for mechanisms with changes in part contact. Numerical kinematic simulation is often used instead. Kinematic simulation takes a driving motion of a part as input and computes motions for the other parts according to the principle that rigid parts cannot overlap. The snapshots in figure 1.3 were generated from a simulation of the indexer mechanism. Rotation of the driver is the driving motion. The lever is assigned a motion that preserves its contacts with the driver and the pin. When the lever tip engages the indexer, the indexer is assigned an angular velocity that prevents tip and tooth overlap.

The drawback of simulation is that it only provides information for one input motion. Unexpected behaviors can be missed. In our example, the simulation assumes that the only driving motion is the driver's rotation. An unexpected force can make the pawl rotate clockwise, disengage from the indexer, and so fail to perform its function. Unexpected behaviors usually involve changes in the contact of parts.

Once the nominal function of the mechanism has been achieved, the next step is to assign tolerances. The goal is to ensure that the kinematic function is preserved for small variations in the shape and configuration of parts. Figure 1.7 shows three examples in which failures occur that are due to small variations of the lever tip. The nominal tip width is 1 mm and its nominal length is 7.5 mm. Too wide a tip (1.8 mm) causes the tip to hit the tooth top (figure 1.7a). Too long a tip (11 mm) causes blocking (b). Too short a tip (5 mm) fails to engage the indexer tooth (c). Tolerance analysis derives the kinematic variation for given tolerances. Synthesis adjusts the parameter values and tolerances to prevent these failures.

Kinematic design can be difficult and time-consuming. The designer has to optimize the design to comply with multiple, conflicting requirements. Many parameter values and tolerances must be adjusted. The adjustments require extensive analysis of many design instances. The analysis involves many contacts of parts, with complicated contact equations. Unintended contacts can arise from variation in parts and can lead to failure modes that coexist with or supersede the nominal function.

A key property of mechanisms is that the interactions of parts are tightly coupled and nonlinear. This property implies that mechanisms cannot be decomposed into linear functional modules with narrow interfaces, whereas other engineering disciplines, notably circuit design, rely heavily on decomposition. The difficulty of mechanism design motivates research in design algorithms.

1.3 Content and Organization of the Book

This book describes our configuration space method for mechanism design, which we have developed over the past 20 years. The presentation is self-contained and tutorial, with references to related work by ourselves and others. It is a research monograph rather than a textbook, so the reader should be prepared to work out some technical details.

The book is organized as follows. In chapters 2–4 we describe our configuration space representation of kinematics. In chapter 2, we review basic geometry concepts, describe the configuration space representation, and discuss kinematic pairs and classification of mechanisms. The shape of a part in defined in terms of geometric features that form its boundary. In chapter 3, we study contacts between two features. We formulate contact equations for basic features, provide closed-form solutions for some cases, and provide general numerical solutions. In chapter 4, we study the configuration spaces of kinematic pairs and describe efficient algorithms for computing them.

In chapters 5–7 we present algorithms for kinematic analysis, tolerancing, and synthesis based on configuration spaces. In chapter 5 we present algorithms for kinematic analysis of a mechanism based on examination of configuration space and kinematic simulation. In chapter 6 we describe tolerance analysis based on a parametric worst-case tolerance model. In chapter 7 we present two methods for kinematic optimization of planar mechanisms.

In chapter 8 we illustrate the configuration space method for designing mechanisms using four case studies taken from industry. The case studies are an optical filter mechanism, a manual transmission gearshift, a MEMS torsional ratcheting actuator, and an automotive asynchronous spatial reverse gear pair. In chapter 9 we conclude with a summary and research directions.

The book contains two appendices. The first is a catalog of 30 representative higher-pair mechanisms. Each mechanism is depicted by a sketch and described with a brief narrative. Its kinematic function is described using configuration spaces. The second appendix contains the users' guide to an open-source C++ software package, HIPAIR, which implements some of the mechanism design methods discussed in this book using configuration spaces.

1.4 Notes

There are numerous introductory and advanced textbooks on kinematic design of mechanisms [22, 31, 59, 78]. Other books focus on linkage [27], cam [4, 26], and gear [53] mechanisms and their design. For an introduction and description of modern mechanical CAD, see Dimarogonas [19].

Commercial packages for mechanical computer-aided design include Catia-CADAM, Pro/Engineer, AutoCAD, I-DEAS, and ADAMS. These packages support drafting, manipulation, and visualization, and include limited support for kinematic analysis and tolerancing of mechanisms whose parts interact via a fixed set of contacts.

Books on robot kinematics include those by Paul [62] and McCarthy [55]. For recent work on robotic manipulation and computational kinematics, see Mason [54] and Angeles et al. [3]. The configuration space method for planning robot motion is described in Latombe [47] and Choset et al. [16].

2 Mechanisms

Mechanisms consist of moving interacting parts. The geometry, motion, and contact of parts give rise to kinematic function. This chapter introduces our representations of these concepts. It describes mechanisms in terms of their basic functional elements, which are kinematic pairs.

In section 2.1 we describe the representation of the geometry of parts. In section 2.2, we describe the configurations (positions and orientations) and motions of parts, and configuration spaces. In sections 2.3 and 2.4 we introduce kinematic pairs and mechanisms and discuss their properties. In section 2.5 we discuss the classification of mechanisms.

2.1 Geometry of a Part

A part in a mechanism is a rigid solid with a fixed shape. Modeling of solids proposes two main types of shape representation: volume (also called constructive solid geometry) and boundary. Volume representations describe parts in terms of set operations (union, intersection, complement) on elementary volumes (cubes, cylinders, cones, spheres). Boundary representations describe parts by their boundary elements. Boundaries are closed surfaces that delimit a part's interior. Volume and boundary representations are mostly interchangeable because it is possible to obtain either one from the other. In this book we use boundary representations because they facilitate the analysis of contacts between parts.

2.1.1 Planar Parts

Although parts are spatial, their kinematic function can often be analyzed in the plane. A planar part is specified by a cross-section and a thickness. For example, the lever from the indexer mechanism in figure 1.2 is a planar part (figure 2.1). The part is obtained by extruding the cross-section perpendicularly to its plane.

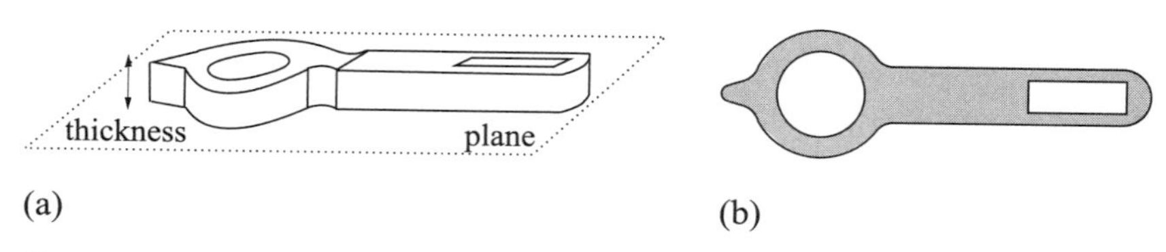

Figure 2.1
A lever (a) and its cross-section (b).

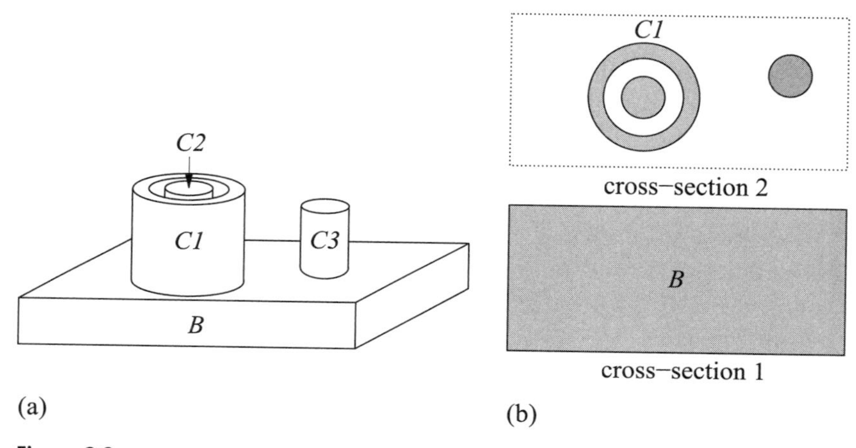

Figure 2.2
A part with two extruded cross-sections (a) and its cross-sections (b).

A cross-section consists of closed regions called components. A component has an outer boundary and may have inner boundaries (holes). Neither components nor boundaries can intersect, but components can nest. The lever has one component with an outer boundary and two holes.

A general planar part consists of a stack of extruded cross-sections with a common plane. Figure 2.2a shows a part with two extruded cross-sections and a nested component. Figure 2.2b shows the two cross-sections. Cross-section 1 has one rectangular component. Cross-section 2 has components $C1$, $C2$, and $C3$, with $C2$ nested inside $C1$.

A boundary is a single closed curve, such as a lever's circular hole, or a sequence of curve segments. Closed curves include circles and ellipses. A curve segment is the portion of a curve delimited by two endpoints. Curve segments include line segments, circle segments, and splines. The outer boundary of the lever consists of line segments and arc segments, whereas the slot boundary has only line segments. The features of a planar

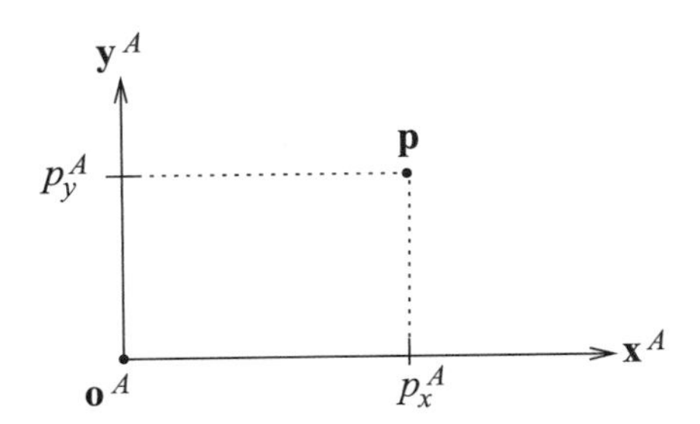

Figure 2.3
Planar part A coordinate system.

part are the closed curves, curve segments, and segment endpoints of its boundary.

Closed curves and curve segments are specified with respect to a part's coordinate system. The coordinate system of a part A consists of a point $\mathbf{o}^A$, called the origin, and of orthogonal unit vectors $\mathbf{x}^A$ and $\mathbf{y}^A$, called axes. The coordinates of point $\mathbf{p}$ in the coordinate system $(\mathbf{o}^A, \mathbf{x}^A, \mathbf{y}^A)$ are denoted by (p_x^A, p_y^A) (figure 2.3).

Curves can be represented explicitly, implicitly, or parametrically. An explicit representation is $y^A = f(x^A)$. An implicit representation is $f(x^A, y^A) = 0$. A parametric representation is $x^A(t) = f(t)$ and $y^A(t) = g(t)$ where t is the curve's parameter. For example, $2x^A + 3y^A + 5 = 0$ is the implicit representation of a line, and $(x^A)^2 + 4(y^A)^2 = 1$ is the implicit representation of an ellipse. We use the implicit representation of curves because it is general and convenient for deriving contact relations.

A boundary is represented by a sequence of adjacent segments. The two endpoints of a segment are labeled tail and head, so that the part's interior is on the left when the segment is traversed from tail to head. Two segments are adjacent when they share an endpoint. The segment sequence is chosen so that the head of each segment is the tail of the next segment. This ordering implies that outer boundaries are traversed counterclockwise, whereas inner boundaries are traversed clockwise. The outer boundary of the lever in figure 2.4 consists of the 12 segments $a_1a_2, a_2a_3, \ldots, a_{12}a_1$ with a_i the segment's endpoints. The rectangular slot boundary consists of the 4 segments c_1, c_2, c_3, c_4, c_1.

In some design tasks we employ a more general, parametric representation of a part. Each boundary segment is specified by a function $f(x, y, \mathbf{p}) = 0$ with $\mathbf{p} = (p_1, \ldots, p_n)$ a parameter vector. The value of the nominal parameter vector defines the part's shape. The shape can be changed by

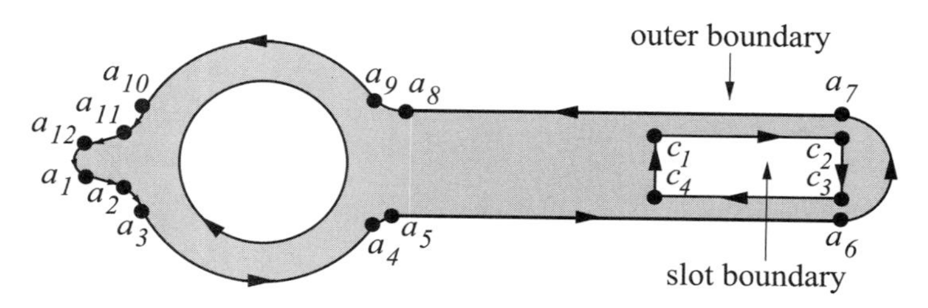

Figure 2.4
Boundary representation of a lever part. The dots represent segment endpoints and arrows indicate segment orientation.

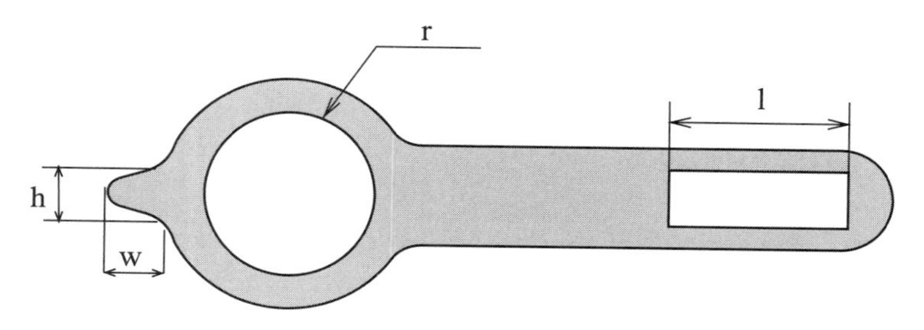

Figure 2.5
Parametric representation of a lever.

changing the parameter's values. The parameters are chosen to reflect shape changes that are relevant to the mechanism's kinematic function.

Figure 2.5 shows a parametric model of the lever. The parameter vector is $\mathbf{p} = (w, h, r, l)$, where w and h are the tip's width and height, r is the the cam's inner radius, and l is the slot length. Its nominal value is $\bar{\mathbf{p}} = (8, 8, 12, 30)$.

2.1.2 Spatial Parts

The boundary representation of a spatial part consists of faces, edges, and vertices, collectively called features, that separate the part's interior from its exterior. Faces are surface patches, edges are line or curve segments where faces meet, and vertices are points where edges meet.

Figure 2.6 shows the most common faces. These faces are obtained by cutting surfaces with planes. For example, the front circle face of the cylinder is the intersection of the plane $y^A = 0$ with the cylinder $(x^A)^2 + (z^A)^2 = 1$. As for planar parts, their topology is defined by the adjacency relations among the features. Parametric spatial models of a part can be defined as for planar parts.

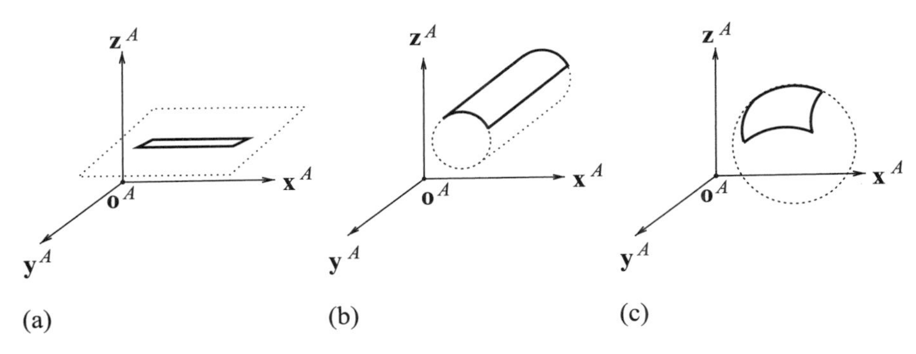

Figure 2.6
Faces: (a) planar, (b) cylindrical, (c) spherical.

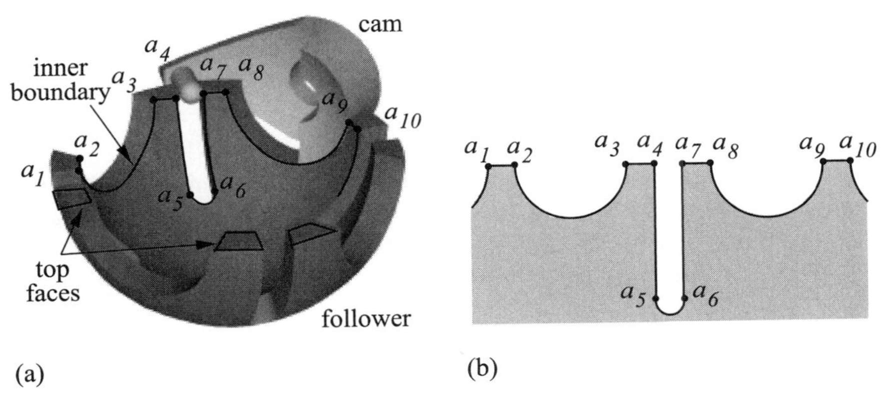

Figure 2.7
Spatial boundary representation: (a) a Geneva pair with part of the follower's inner spherical patch boundary highlighted, (b) flattened projection of part of the inner spherical patch follower's boundary profile. Dots represent vertices.

The spatial Geneva pair illustrates these concepts (figure 2.7). The cam consists of a plate with a cylindrical pin and a half cylinder mounted on it. The follower is a hollow hemisphere with four evenly spaced slots and circular cutouts. The boundary representation consists of spherical and cylindrical faces bounded by line and arc segments. The follower's inner boundary consists of a spherical patch with four slots and four cutouts. The boundary of each slot consists of two parallel circle segments on the sides and a semicircle at the bottom. The slot's boundary and the circle segment's cutout are connected by a circle segment. The pattern repeats four times to form a closed boundary. Figure 2.7b shows a flattened projection of part of this boundary. The vertices a_i correspond to boundary segment endpoints: segment a_1a_2 is a straight line, followed by a circle segment a_2a_3 from the cutout, connected by line segment a_3a_4 to the slot boundary consisting of arc segments a_4a_5 (which projects to a line segment), a_5a_6, and a_6a_7. The follower's outer boundary has the same structure, but with larger dimensions. Both are connected by eight rectangular planar faces at the top, as shown in figure 2.7a.

2.2 Configurations and Motions of Parts

This section describes the mathematics of the configurations and motions of rigid parts. The configuration of a part is its position and orientation in a world coordinate system. Its position is given by the world coordinates of the part's origin. The orientation is the angles between the part's axes and the world axes. The part's configuration determines the set of world points that it occupies. We begin with planar parts and then extend the discussion to spatial parts.

2.2.1 Planar Parts

The configurations of parts are specified using two coordinate systems as shown in figure 2.8a. The world coordinate system, $(\mathbf{o}, \mathbf{x}, \mathbf{y})$, is fixed; for example, it is attached to the mechanism's frame. The part A coordinate system, $(\mathbf{o}^A, \mathbf{x}^A, \mathbf{y}^A)$, is attached to the part. The part's position is specified by the translation vector $\mathbf{t}_A = (x_A, y_A)$. The part's orientation is specified by the rotation matrix

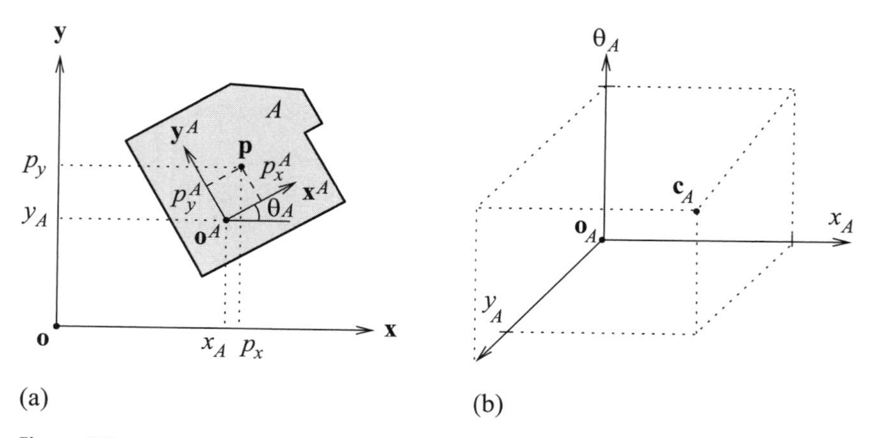

Figure 2.8
Planar part coordinate systems: (a) world and part coordinate system, (b) a configuration space coordinate system.

$$R_A = \begin{bmatrix} \cos\theta_A & -\sin\theta_A \\ \sin\theta_A & \cos\theta_A \end{bmatrix}$$

that maps the part's axes to the world axes. Part coordinates, $\mathbf{p}^A = (p_x^A, p_y^A)$, are mapped to world coordinates, $\mathbf{p} = (p_x, p_y)$, by the Euclidean transformation

$$\mathbf{p} = \mathbf{t}_A + R_A \mathbf{p}^A.$$

The variables x_A, y_A, and θ_A are called configuration variables. Configuration variables x_A and y_A are called position variables and θ_A is an orientation variable. The domain of x_A and y_A is the real line, $\Re$, and the domain of θ_A is the unit circle S ($-180^\circ \le \theta_A \le +180^\circ$). Their Cartesian product, $\Re^2 \times S$, is called configuration space, and its points, $\mathbf{c}_A = (x_A, y_A, \theta_A)$, are called configurations. The configuration space represents the set of all positions and orientations (figure 2.8b) of a part.

Figure 2.9 shows a planar part in configurations $\mathbf{c}_1$, $\mathbf{c}_2$, $\mathbf{c}_3$, $\mathbf{c}_4$. In configuration $\mathbf{c}_1 = (2, 2, 0^\circ)$, the part's origin is at two horizontal and vertical units (e.g., centimeters) from the world origin and their axes are aligned. In configuration $\mathbf{c}_2 = (11, 2, 0^\circ)$, the part's origin is moved by nine units along the $\mathbf{x}$ axis. In configuration $\mathbf{c}_3 = (16, 8, 0^\circ)$, the part's origin is moved by five units further along the $\mathbf{x}$ axis and by six units along the $\mathbf{y}$ axis. In configuration $\mathbf{c}_4 = (16, 8, 30^\circ)$, the part is rotated counterclockwise by 30°. Figure 2.10 shows the four configurations in configuration space (x_A, y_A, θ_A).

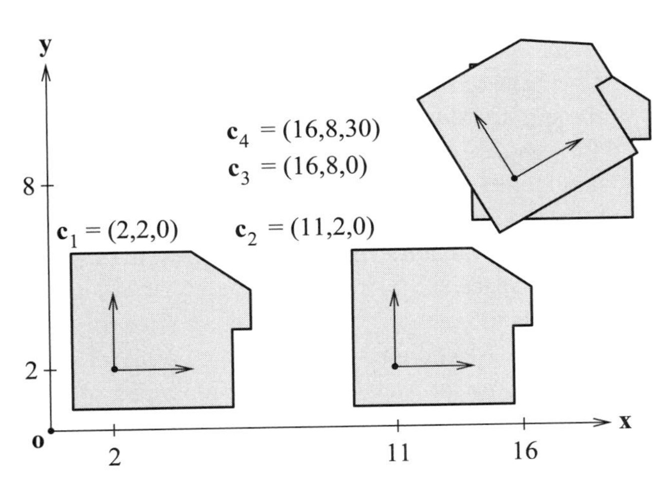

Figure 2.9
A part in configurations $\mathbf{c}_1$, $\mathbf{c}_2$, $\mathbf{c}_3$, $\mathbf{c}_4$ in the plane.

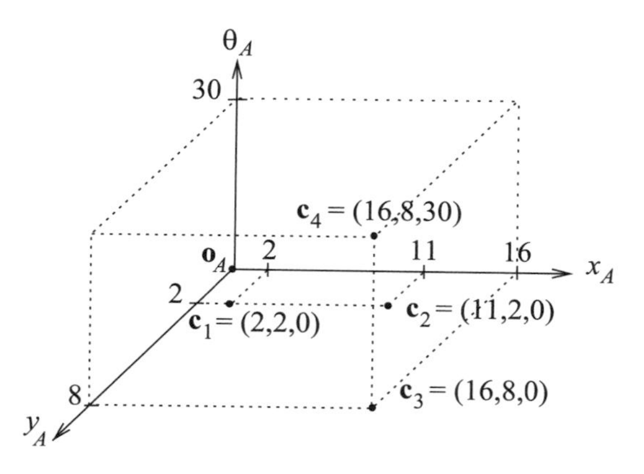

Figure 2.10
Part configurations $\mathbf{c}_1$, $\mathbf{c}_2$, $\mathbf{c}_3$, $\mathbf{c}_4$ in configuration space.

When a part moves, its configuration changes continuously over a time interval. The part's configuration is $\mathbf{c}_A(t) = (x_A(t), y_A(t), \theta_A(t))$ where $x_A(t)$, $y_A(t)$, and $\theta_A(t)$ are the values of x_A, y_A, and θ_A at time t. The changing configuration traces a path in configuration space. When the position variables are constant and the orientation variable changes, the motion is rotation around a fixed axis. When the orientation variable and one position variable are constant, the motion is translation parallel to the vertical or horizontal axis. When the position variables are linearly related, the motion is translation along a fixed axis defined by the linear relation. Parts with these motions are called fixed-axis parts.

Figure 2.11 shows an example of a planar moving part. The motion starts at configuration $\mathbf{c}_1$ and translates along the $\mathbf{x}$ axis nine units relative to $\mathbf{c}_1$ to reach $\mathbf{c}_2$. It then translates along a diagonal to reach $\mathbf{c}_3$ and rotates by 30° to reach the final configuration, $\mathbf{c}_4$. The part's motion thus consists of three stages: horizontal translation, diagonal translation, and rotation. Figure 2.12 shows the configuration variables' time plots and figure 2.13 shows the path in the configuration space.

In many mechanisms, parts are designed to move along fixed axes. To simplify the kinematic analysis, we choose a part coordinate system so that two of the three configuration variables are constant. When a part rotates, the coordinate system's origin is placed at the center of rotation, so that the two position variables are constant. When a part translates, the coordinate system's $\mathbf{x}$ axis is aligned with the motion axis so that the $\mathbf{y}$ variable and the orientation variable are constant. For example, suppose part A rotates

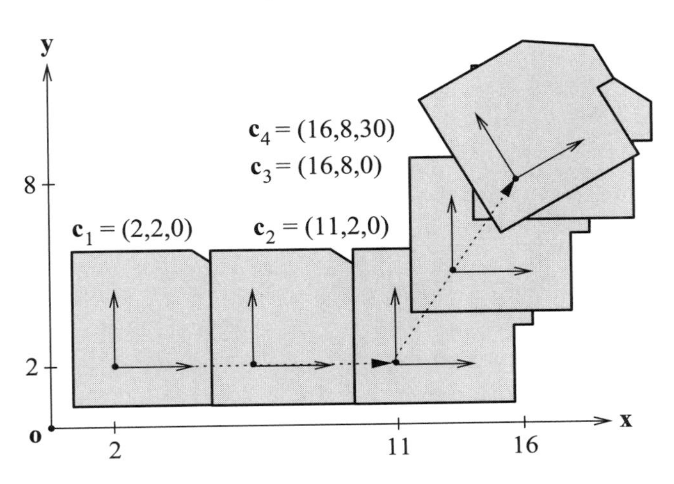

Figure 2.11
Representative motion snapshots of the planar motion of a part.

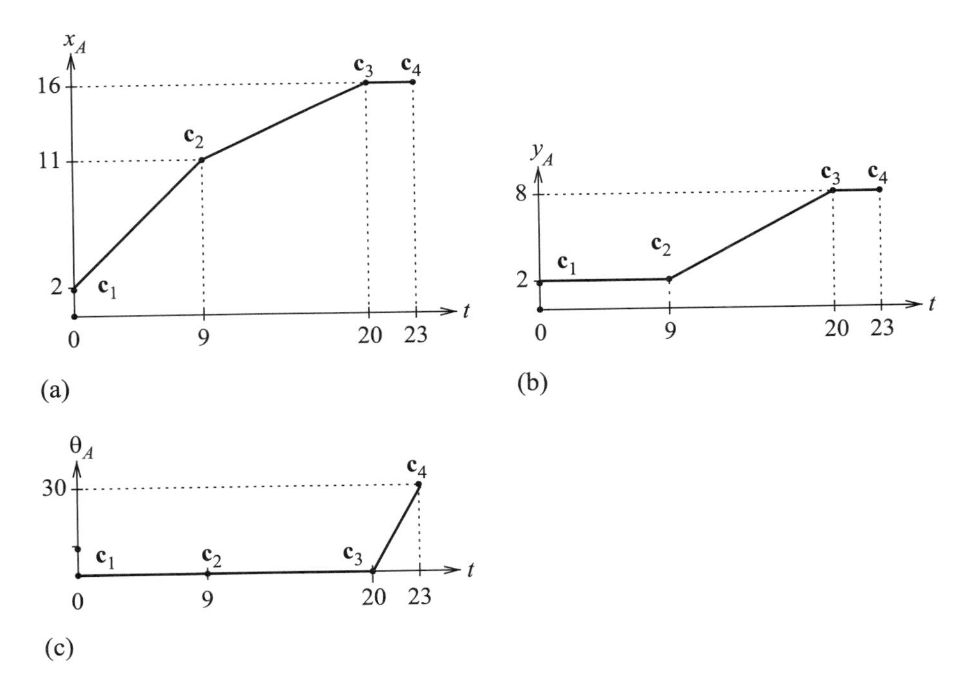

Figure 2.12
Planar motion time plots: (a) $x_A(t)$, (b) $y_A(t)$, (c) $\theta_A(t)$.

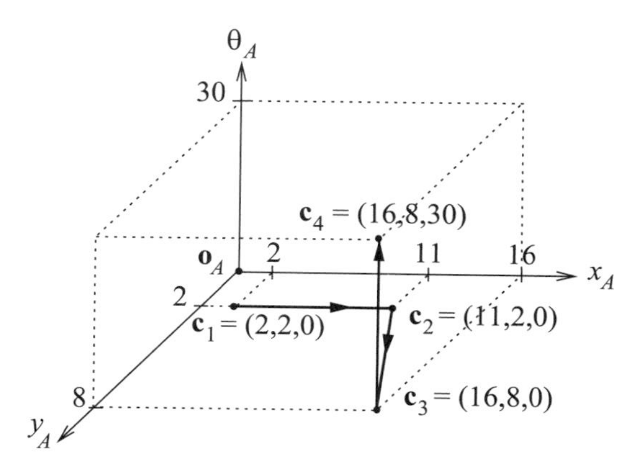

Figure 2.13
Planar motion of a part in configuration space.

around its origin $\mathbf{o}_A$ in figure 2.8. The part's configuration (x_A, y_A, θ_A) is $(3, 2, \theta_A)$, which simplifies to θ_A.

The independent variables that specify the configuration of a part are called the degrees of freedom of the part. The number of variables is called the number of degrees of freedom. Fixed-axis planar parts have one degree of freedom whereas free planar parts have three.

2.2.2 Spatial Parts

Spatial configurations of parts are analogous to planar ones (figure 2.14). The world coordinates are $(\mathbf{o}, \mathbf{x}, \mathbf{y}, \mathbf{z})$ and the part A coordinates are $(\mathbf{o}^A, \mathbf{x}^A, \mathbf{y}^A, \mathbf{z}^A)$. A spatial configuration consists of six variables, versus three in the plane. The translation vector is $\mathbf{t}_A = (x_A, y_A, z_A)$. The three-by-three rotation matrix, R_A, is expressed in terms of three orientation variables. One option, called Euler angles, is to express a general rotation matrix as the product of three rotations around coordinate axes, each of which has a one-parameter rotation matrix. In the roll-pitch-yaw convention (figure 2.14b), the inverse of R_A is expressed as

$$\begin{bmatrix} \cos\theta_A & -\sin\theta_A & 0 \\ \sin\theta_A & \cos\theta_A & 0 \\ 0 & 0 & 1 \end{bmatrix} \times \begin{bmatrix} \cos\phi_A & 0 & -\sin\phi_A \\ 0 & 1 & 0 \\ -\sin\phi_A & 0 & \cos\phi_A \end{bmatrix} \times \begin{bmatrix} 1 & 0 & 0 \\ 0 & \cos\psi_A & -\sin\psi_A \\ 0 & \sin\psi_A & \cos\psi_A \end{bmatrix}$$

with θ_A the roll angle around x_A, ϕ_A the pitch angle around y_A, and ψ_A the yaw angle around z_A. The transpose of this product is R_A.

The configuration space of a spatial part is six-dimensional, with position variables x_A, y_A, z_A and orientation variables θ_A, ϕ_A, ψ_A. The position vari-

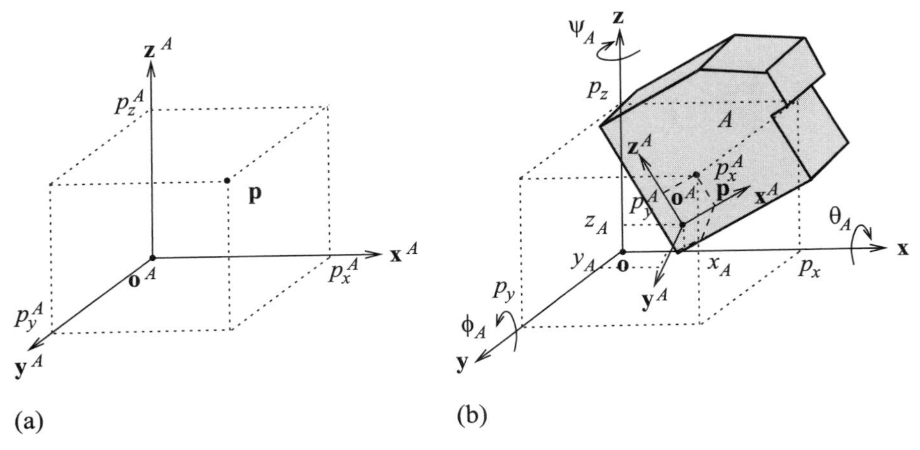

Figure 2.14
Spatial coordinates system and configuration of a part: (a) world and parts coordinate system, (b) configuration of a part.

able domain is the real line and the orientation variable domain is [$-180°$, $+180°$]. As in the plane, spatial motions specify configurations of a part as a continuous function of time. We only study spatial parts that rotate or translate around fixed axes.

2.3 Kinematic Pairs

The basic functional element of a mechanism is the kinematic pair. A kinematic pair is formed by two parts in contact that move relative to each other. We first describe the configuration space representation for planar and spatial pairs and then discuss lower and higher pairs.

2.3.1 Representation of Configuration Space

We represent kinematic pair configurations in a configuration space. We consider the cases of planar pairs and spatial pairs. For the planar case, let A and B be two planar parts with configuration variables x_A, y_A, θ_A and x_B, y_B, θ_B as shown in figure 2.15. The configuration variables define a six-dimensional space, called the pair configuration space, which is the Cartesian product of the parts' configuration spaces. A point in this space defines the configuration of the two parts. The parts' configuration variables become coordinates in the configuration space. For example, in figure 2.15, part A is in configuration $(x_A, y_A, \theta_A) = (3, 2, 30°)$ and part B is in configuration $(x_B, y_B, \theta_B) = (7, 1.75, 45°)$, so the configuration

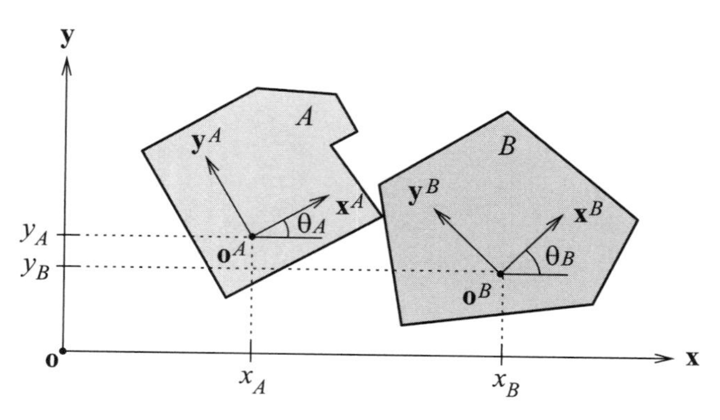

Figure 2.15
Coordinate systems and configuration variables for a part.

$$(x_A, y_A, \theta_A, x_B, y_B, \theta_B) = (3, 2, 30°, 7, 1.75, 45°)$$

is a contact configuration.

We consider the cases of fixed-axis pairs and general pairs. In a fixed-axis pair, four of the six configuration variables of the pair are constant. For example, in figure 2.15, suppose part A rotates around its origin $\mathbf{o}_A$ and part B translates along its $\mathbf{x}_B$ axis. The pair configuration, $(x_A, y_A, \theta_A, x_B, y_B, \theta_B)$ is $(3, 2, \theta_A, x_B, 1.75, 45°)$, so it can be simplified to (θ_A, x_B), which defines a two-dimensional configuration space. There are three types of two-dimensional configuration spaces for planar fixed-axis pairs: rotation-rotation with configuration space coordinates θ_A and θ_B; rotation-translation with configuration space coordinates θ_A and x_B, x_A, and θ_B; and translation-translation with configuration space coordinates x_A and x_B.

In a general planar pair, each part has three degrees of freedom. Instead of describing pair configurations in the world coordinate system, we describe the configuration of moving part A relative to fixed part B (figure 2.16). Part B is fixed at the world coordinate system's origin with its axes aligned with the world axes, so its configuration is $(0, 0, 0°)$. Part A configuration variables x_A, y_A, θ_A become the coordinates of A relative to B. The pair configuration $(x_A, y_A, \theta_A, x_B, y_B, \theta_B)$ is replaced by (x_A, y_A, θ_A), which defines a three-dimensional configuration space.

For the spatial case, let A and B be two spatial parts with configuration variables x_A, y_A, z_A, θ_A, θ_A, ϕ_A, ψ_A and x_B, y_B, z_B, θ_B, θ_B, ϕ_B, ψ_B. The configuration variables define a 12-dimensional configuration space. In a fixed-axis pair, each part has one degree of freedom and we choose coordinate systems in which four of the six configuration variables are constant. In a general pair, we can employ relative coordinates.

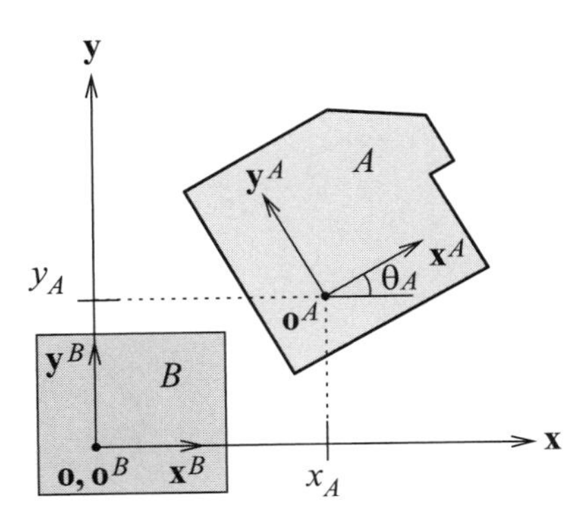

Figure 2.16
Relative coordinates for general planar pair.

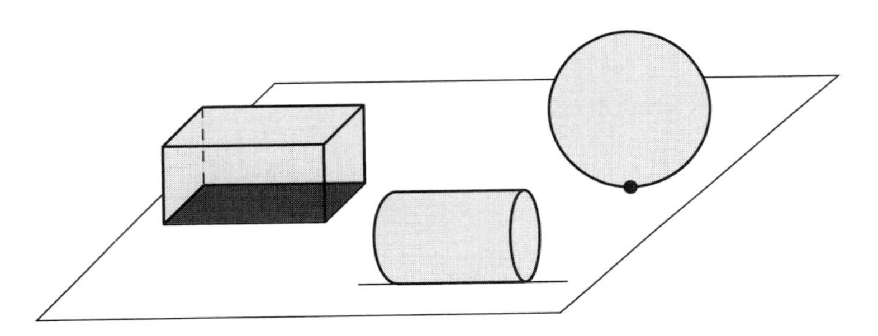

Figure 2.17
Contact types.

Contacts between parts constrain their configurations by inducing functional relations among their configuration variables. For example, assume that the table in figure 2.17 is fixed, with the world coordinate frame chosen so that the **x** and **y** coordinates lie on the table's surface. The sphere A can only be placed so that the distance from its center to the table is equal to the sphere's radius. The sphere's center lies on a plane parallel to the table's surface, so its z_A position variable is constant. The variables of the block position are similarly constrained, and two of its three orientation variables are constant. In a pair with permanent contacts of parts, the configuration can be specified by an independent subset of the configuration variables, called the degrees of freedom of the pair. For example, the sphere-table pair has five degrees of freedom, the block-table pair has three, and an involute gear pair has one.

The parametric representation of a kinematic pair is a direct extension of the parametric representation of its parts. The parameter vector **p** defines

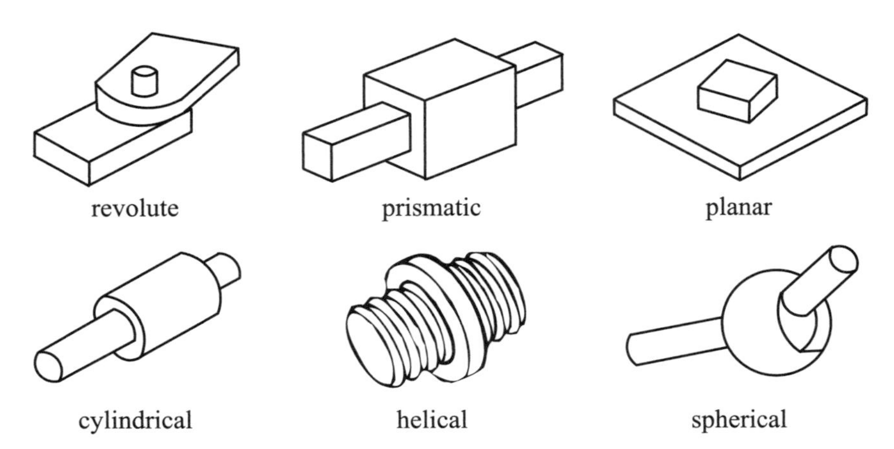

Figure 2.18
The six kinematic lower pairs.

the parts' shapes and motion axes, and $\bar{\mathbf{p}}$ specifies the nominal parameter values.

2.3.2 Lower and Higher Pairs

A contact occurs when a feature of one part touches a feature of another part. Contacts can be permanent or changing. An example of permanent contact is a cylindrical shaft mounted in a hub with a matching hole. An example of changing contacts is the two transmission plates of a clutch. Also, two parts can have multiple contacts, such as two rotating parallel gears with several meshing teeth.

The contact region of two parts is the set of points at which two features touch. There are three types of contact regions: surfaces, lines, and points. Figure 2.17 illustrates the three cases: a block on a table is a surface contact, a cylinder on a table is a line contact, and a sphere on a table is a point contact.

The mechanical engineering literature classifies kinematic pairs as lower and higher. Lower pairs involve a single, permanent surface contact between the features of two parts. Higher pairs are all other pairs, including multiple contacts, changing contacts, line contacts, and point contacts.

Lower pairs, also called joints, are the most popular and robust. There are six types of lower pairs: three planar and three spatial (figure 2.18). Planar lower pairs contain parts whose surface contacts induce planar motion: revolute, prismatic, and planar, also called pin, slider, and plane joints. Spatial lower pairs contain parts whose surface contacts induce motions in space:

cylindrical, helical, and spherical, also called cylinder, screw, and sphere joints. Revolute and prismatic pairs have one degree of freedom: rotation or translation along a fixed axis. Plane pairs have all three planar degrees of freedom. Helical pairs have one degree of freedom: linearly coupled rotation and translation along a fixed axis. Cylindrical pairs have independent rotation and translation along a fixed axis. Spherical pairs have three independent rotations around a fixed point.

Higher pairs can be further classified by their motion types: fixed-axis or general, and planar or spatial. Planar pairs consist of two planar parts with a common plane whose motion is parallel to that plane. Planar fixed-axis pairs have one degree of freedom per part. General planar pairs contain parts with planar motions and three degrees of freedom, whereas general spatial pairs contain parts with spatial motions and up to six degrees of freedom.

Figure 2.19 shows four examples of kinematic pairs. The examples use thick-headed arrows to indicate intended motion directions, as explained in section 1.1. Figure 2.19a shows a sector gear pair, which is a fixed-axis planar pair. Each gear rotates around its center on a fixed axis. The continuous rotation of the driver intermittently turns the follower half a turn in the opposite direction. Figure 2.19b shows a general planar pair. The gear and driver cam are fixed-axis parts, and the enclosed finger cam rotates and translates in the plane. Figure 2.19c is a spatial fixed-axis pair in which the worm gear and the follower gear rotate around fixed axes that are orthogonal to each other. The spatial geometry of the worm screw and the gear teeth makes the pair spatial rather than planar. Figure 2.19d shows a ball-bearing ratchet mechanism. When the driver rotates counterclockwise around the fixed axis at its center, it drives the follower clockwise by means of four ball bearings, which form general spatial pairs with the driver and the follower. Motion in the clockwise direction is prevented by the shape of the cam's inner notches.

2.4 Mechanisms

Mechanisms consist of interacting fixed and moving parts. The fixed parts, called the frame, serve as a support for the moving parts. The moving parts interact among themselves and with the frame via contacts. Their shapes and interactions are designed to achieve useful kinematic functions.

As for kinematic pairs, we represent a mechanism's configurations in configuration space. In a planar mechanism with n moving parts, each part contributes up to three configuration variables, so the configuration

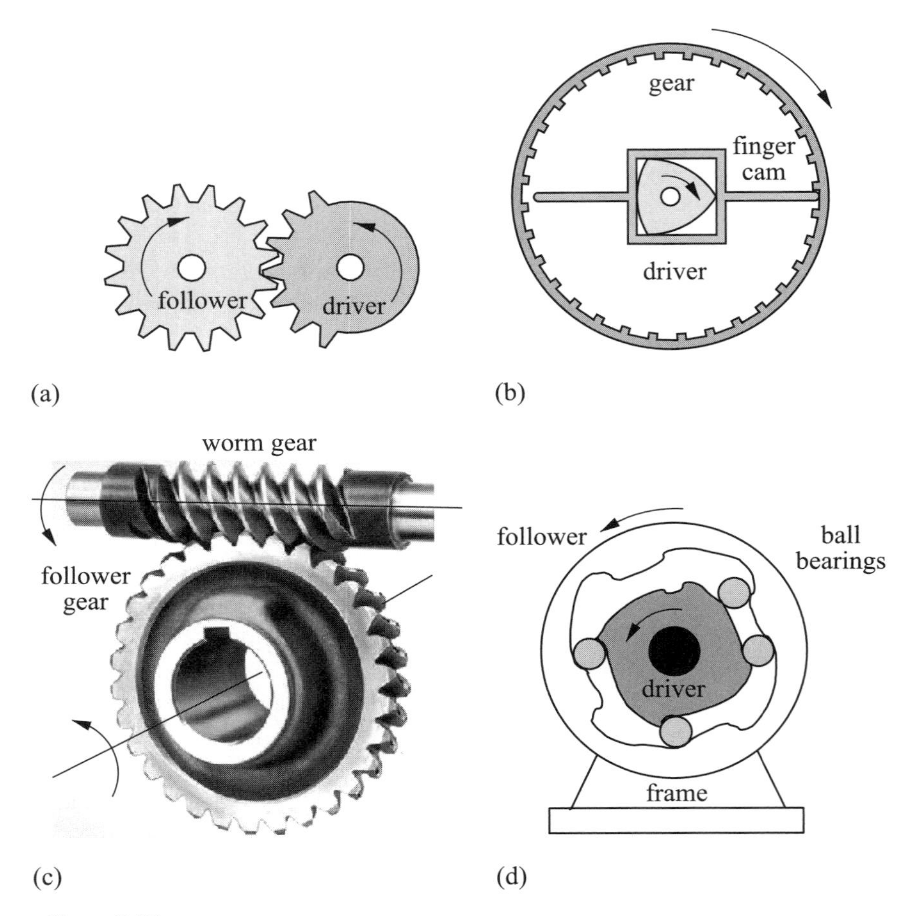

Figure 2.19
Examples of kinematic higher pairs: (a) planar fixed-axis, (b) general planar, (c) spatial fixed-axis, (d) general spatial.

space is at most $3n$-dimensional. In a spatial mechanism, each moving part contributes up to six configuration variables, so the configuration space is at most $6n$-dimensional.

Kinematic functions describe the relationships among the motions of parts. For example, the function of two parallel fixed-axis gears with meshing teeth of equal diameter is to reverse the rotation direction while keeping the same rotation speed and acceleration. The kinematic functions result from the shapes, contacts, and driving motions of the parts.

Mechanisms are most often described by their structural and functional properties. Structural properties describe the shapes, features, relative con-

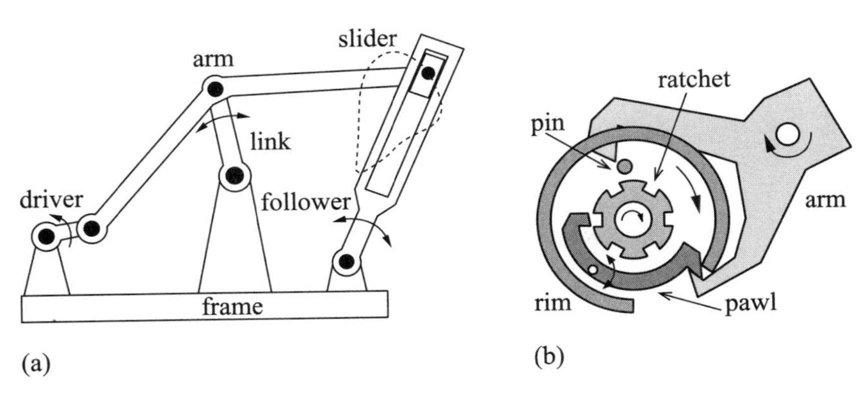

Figure 2.20
Planar mechanisms: (a) link-gear dwell, (b) Dennis clutch.

figurations, and contacts of parts. For example, the structural description of a gearbox indicates the number of gears, their rotation axes, number of teeth and diameter, and the contacts between the gears. Additional structural descriptions refer to geometric properties common to all parts and pairs. Planar mechanisms include planar parts and pairs with a common plane, whereas spatial mechanisms include one or more spatial pairs. Fixed-axis mechanisms consist of parts that move along fixed axes. Linkage mechanisms consist of lower pairs only. Other structural descriptions stem from the common name of the main part or pair of the mechanism, such as gear, screw, piston, cam, gear-crank, and lever-ratchet.

The mechanisms in figure 2.20 illustrate these concepts. The link-gear dwell mechanism is a planar linkage consisting of a fixed frame and five moving parts with six pin joints (solid circles) and one slider joint. The counterclockwise rotation of the driver causes the follower to oscillate and the slider to follow the closed trajectory (dotted lines). The Dennis clutch is a general planar mechanism consisting of an arm, a ratchet, a cam, a pawl, and a frame (not shown). The arm is attached to the frame with a pin joint. The ratchet is mounted on a rotating shaft. The cam consists of a pin and a rim attached to a plate, and is mounted on a rotating shaft. The pawl is attached to the cam plate by a pin. Figure 2.7a shows a fixed-axis spatial mechanism consisting of a cam and a follower mounted on a frame (not shown).

Contacts between parts are collectively referred to as the mechanism's contact topology. An abstract description of the contacts is a topology graph. In this graph, each part is represented by a node and an edge between two nodes represents a contact between two parts. Solid edges represent permanent contacts, while dashed edges represent intermittent

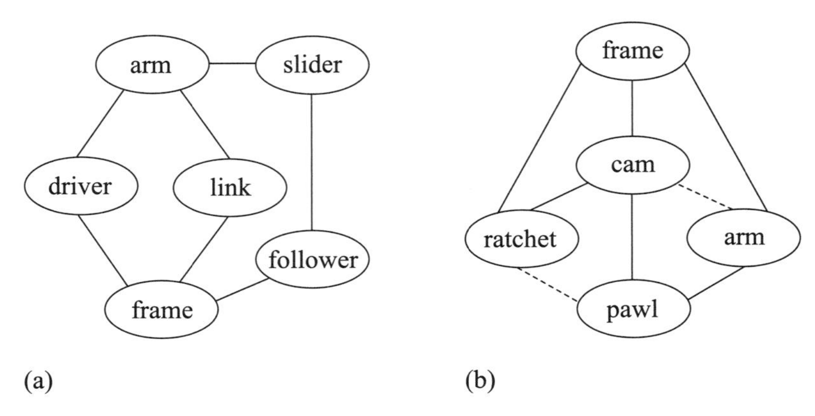

Figure 2.21
Topology graphs of the mechanisms in figure 2.20: (a) link-gear dwell and (b) Dennis clutch.

contacts. Paths between nodes in the graph indicate interactions between parts that are not directly in contact. A loop in the graph is a path from a part to itself that goes through two or more parts. The structure of the graph serves to characterize mechanisms. Topology graphs can be open loop and closed loop, depending on whether they contain a loop. Open-loop graphs can be trees or chains. A tree is a graph in which each node except one, the start node, has exactly one parent and zero or more children. A chain is a tree in which each node but one, the end node, has exactly one child. Closed-loop graphs can have a single or multiple loops. Figure 2.21 shows the topology graphs of the mechanisms in figure 2.20.

Functional descriptions refer to the motions of parts and their relationships. These motions can be simple, such as fixed-axis rotation or translation, or more complex. Motion relations describe the relationships between a part's degrees of freedom in terms of mathematical functions. For example, the functional description of the gearbox indicates the gear shifts and the ratio between the input and output rotation axes. Additional functional descriptions refer to the overall purpose of the mechanism. Examples include clutch, braking, indexing, stopping, and locking mechanisms. Often, motions of parts are interpreted as input and output motions.

The two mechanisms in figure 2.20 illustrate these concepts. The function of the link-gear dwell mechanism is to convert a continuous rotation of the driver to an oscillatory motion of the follower, with a dwell in-between. The function of the Dennis clutch is to prevent the clockwise rotation of the cam.

2.5 Classification of Mechanisms

It has long been recognized that the classification of mechanisms according to their structural and functional properties poses a difficult problem. The mechanical engineering literature often employs structural, functional, and other descriptions, such as the inventor's name (Dennis clutch) and the invention location (Geneva cam).

The literature includes a variety of collections and encyclopedias on mechanisms, loosely organized by structure and function. The most comprehensive and systematic collection is Artobolevsky's *Mechanisms in Modern Engineering Design* [5]. The five-volume set contains 4,745 mechanisms classified in a two-tier group system. The first classification is based on structural features such as lever, cam, gear, link-gear, and slider-crank, to name a few. The second classification is based on their function: balance, brake, dwell, guiding, etc. The drawings have been simplified to retain only the relevant shape details and are supplemented with short explanations. Figure 2.22 shows two entries from the encyclopedia. The four-volume collection *Ingenious Mechanisms for Designers and Inventors* edited by Horton and Jones [30] includes hundreds of mechanisms loosely classified by mechanical motion, such as tripping and stop, intermittent motion, and differential motion.

Since the focus of this book is computational, we elaborated our own classification of mechanisms according to their geometric and kinematic properties. We conducted a survey of about 2,500 rigid-part mechanisms in Artobolevsky's encyclopedia to identify the main categories and determine their relative importance. We excluded mechanisms with flexible parts, such as belts and chains because they fall outside the scope of standard kinematics.

Tables 2.1 and 2.2 summarize the classification results for higher pairs and for mechanisms. Of the 559 stand-alone kinematic higher pairs, 72% are planar and 28% are spatial; 89% have one or two degrees of freedom while 11% have three or more. Most pairs (66%) are both planar and fixed-axis. The 1,912 mechanisms surveyed have between 3 and 25 moving parts, and 7 on average. Linkages comprise 35% of all mechanisms, fixed-axis 22%, and fixed-axis parts connected by linkages 9%. A total of 85% mechanisms are planar and 18% have changing contacts (such as clutch and ratchet mechanisms).

The mechanism compilations and our survey show that higher pairs are pervasive. The most common are gear and cam pairs, which appear in all types of mechanisms. Others, such as ratchet, escapement, and indexer pairs, are common in low-torque mechanisms, such as sewing machines,

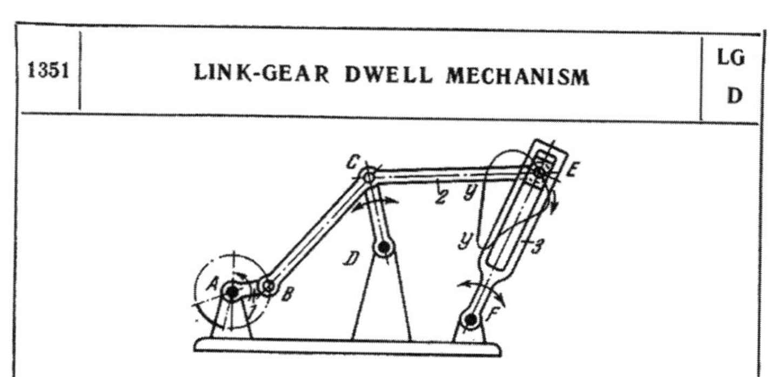

1351 | **LINK-GEAR DWELL MECHANISM** | LG D

The lengths of the links comply with the conditions: $\overline{BC} = 4.28\overline{AB}$, $\overline{CE} = 4.86\overline{AB}$, $\overline{BE} = 8.4\overline{AB}$, $\overline{CD} = 2.14\overline{AB}$, $\overline{AD} = 4.55\overline{AB}$, $\overline{AF} = 7\overline{AB}$ and $\overline{DF} = 3.32\overline{AB}$. When point B of crank *1* travels along the part of a circle indicated by a heavy continuous line, point E of connecting rod *2* describes portion y-y of its path that is shown by a heavy continuous line and approximates a straight line passing through point F. During continuous rotation of crank *1* about fixed axis A, slotted link *3* oscillates about fixed axis F with a dwell during the travel of point E along portion y-y of its path.

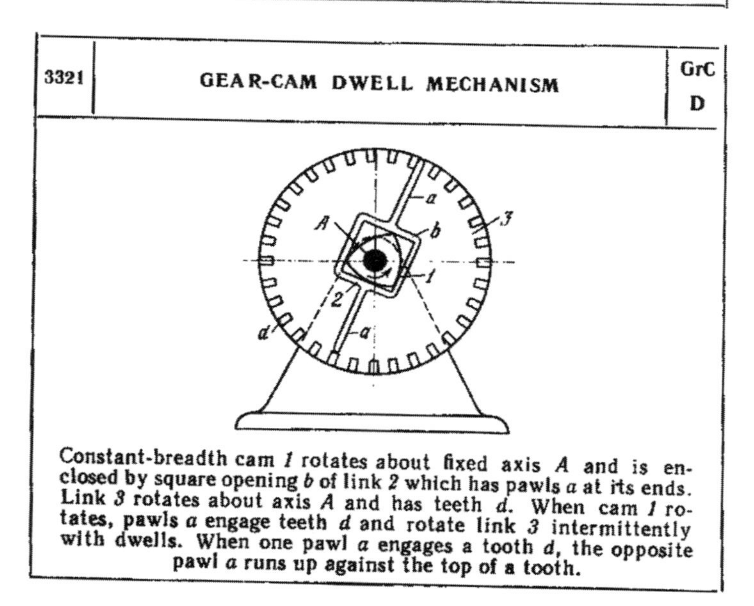

3321 | **GEAR-CAM DWELL MECHANISM** | GrC D

Constant-breadth cam *1* rotates about fixed axis A and is enclosed by square opening b of link *2* which has pawls a at its ends. Link *3* rotates about axis A and has teeth d. When cam *1* rotates, pawls a engage teeth d and rotate link *3* intermittently with dwells. When one pawl a engages a tooth d, the opposite pawl a runs up against the top of a tooth.

Figure 2.22

Two examples of entries from Artobolevsky's *Mechanisms in Modern Engineering Design*.

Table 2.1
Survey of kinematic pairs

		Degrees of freedom		Geometry		
Volume	Total	≤2	>2	planar	spatial	Planar fixed-axis
Lever mechanisms 1	159	130	29	102	57	94
Lever mechanisms 2	56	43	13	46	10	37
Gear mechanisms	180	174	6	140	40	136
Cam mechanisms	164	148	16	114	50	103
Total	559	495	64	402	157	370
Percentage	—	89%	11%	72%	28%	66%

Notes: The first column shows the total number of pairs. The following two columns show the number of pairs with two or fewer degrees of freedom, or with more than two. The fourth and fifth columns indicate the number of planar and spatial pairs. The last column indicates the number of fixed-axis pairs.

Table 2.2
Survey of Mechanisms

		Motion type					
Volume	Total	Linkage	FA	Both	Others	VC	Spatial
Lever mechanisms 1	614	341	101	18	154	123	108
Lever mechanisms 2	677	323	105	66	183	110	70
Gear mechanisms	350	1	106	36	207	72	28
Cam mechanisms	271	4	101	47	119	40	72
Total	1912	669	413	167	663	340	278
Percentage	—	35%	22%	9%	35%	18%	15%

Notes: The first column shows the total number of mechanisms. The following three columns show the number of linkage, fixed-axes (FA), and linkage and fixed-axes (Both) mechanisms. The fifth columns shows the number of remaining mechanisms. The sixth column shows the number of mechanisms with varying contacts (VC). The last column shows the number of spatial mechanisms.

copiers, cameras, and VCRs. Mechanisms with one degree of freedom, such as the link-gear dwell mechanism, are the most common.

2.6 Notes

Representation of the shape of parts has been the topic of extensive research in the solid modeling community for the past 30 years. An excellent textbook on the subject is by Mantyla [53].

The motions of parts are studied in classic mechanical engineering and robotics literature. There are textbooks by Erdmann and Sandor [22] and Paul [62].

The configuration space representation of locations and motions of parts is presented in Latombe [47], and more recently in Choset et al. [16].

Collections and encyclopedias describing many useful and practical mechanisms include those by Artobolevsky [5], Horton and Jones [30], Chironis [15], and Jensen [33]. More details on the survey we conducted on Artobolevsky's encyclopedia can be found in Joskowicz and Sacks [40]. Tsai [79] addresses classification of mechanisms based on graph theory to characterize the structure and function for mechanisms with permanent contacts.

The KMODDL on-line Cornell University Reuleaux Collection of Kinematic Mechanisms includes many historical texts, mechanisms, and their animations at `http://kmoddl.library.cornell.edu`.

Appendix B describes the representation of parts and mechanisms in HIPAIR. It covers fixed-axis planar parts whose boundaries are line and circle segments.

3 Contact of Features

This chapter begins the study of contacts of parts with the concept of contacts of features. Two features are in contact when they share one or more points, called contact points, and do not intersect. We study the contact of features in the configuration space of a kinematic pair. We compute the contact configurations by formulating and solving contact equations. We solve the equations in closed form when possible and use a numerical solver in other cases. Using a closed-form solution is faster and more accurate than using a numerical solver, but is less general.

The simplest planar features are line segments, circle segments, and their endpoints, called vertices. These are the most common features in planar kinematic pairs. In the first three sections, we formulate their contact equations and derive closed-form solutions. We formulate contact equations for general planar features in section 3.4 and for spatial features in section 3.5. Few cases have closed-form solutions, so we turn to numerical solutions.

3.1 Simple Planar Feature Contact

There are six pairs of simple planar features: circle-circle, circle-line, line-line, vertex-vertex, circle-vertex, and line-vertex. Each pair defines a contact type. In this section we derive contact equations for the first two types and reduce the other four types to these two. We solve the contact equations for fixed-axis pairs in section 3.2 and for general planar pairs in section 3.3.

Two segments are in contact when the underlying circles or lines are tangent (figure 3.1a) and the point of tangency lies on both segments (figure 3.1b). The tangency configurations form the solution set of a circle-line or circle-circle tangency equation. The contact configurations are a subset bounded by the configurations where the point of tangency coincides with a vertex (figure 3.1c). The boundary configurations are the solutions of vertex-line and vertex-circle boundary equations.

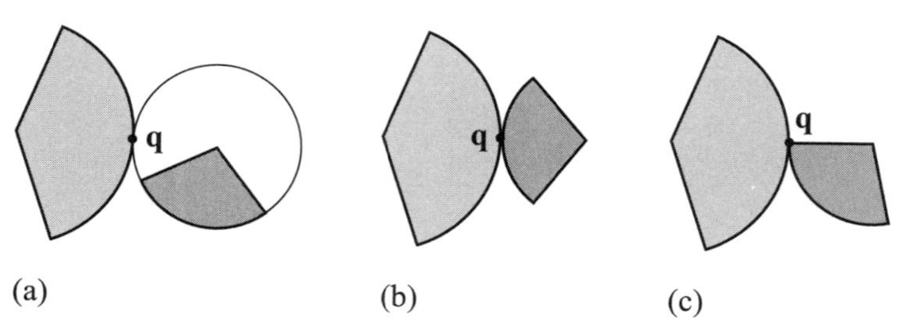

Figure 3.1
Tangency (a), contact (b), and boundary configuration (c) at $\mathbf{q}$.

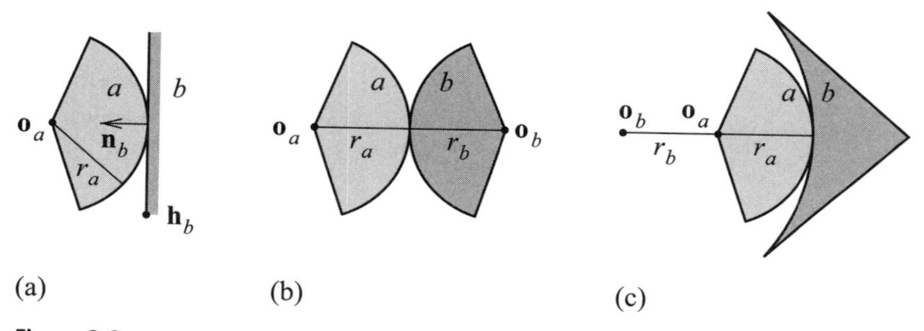

Figure 3.2
Tangency conditions: (a) circle-line, (b–c) circle-circle.

3.1.1 Tangency Equations

Figure 3.2a shows the tangency condition for a circle, a, with radius r_a and center $\mathbf{o}_a$, and a line, b, with point $\mathbf{h}_b$ and unit normal $\mathbf{n}_b$. The distance from the circle's center to the line equals the circle's radius: $\mathbf{n}_b \cdot (\mathbf{o}_a - \mathbf{h}_b) = r_a$ where the dot denotes the inner product $\mathbf{v} \cdot \mathbf{w} = v_x w_x + v_y w_y$. We express this equation in the coordinates of parts by substituting

$$\mathbf{o}_a = R_A \mathbf{o}_a^A + \mathbf{t}_A$$

$$\mathbf{h}_b = R_B \mathbf{h}_b^B + \mathbf{t}_B$$

$$\mathbf{n}_b = R_B \mathbf{n}_b^B$$

to obtain

$$R_B \mathbf{n}_b^B \cdot (R_A \mathbf{o}_a^A + \mathbf{t}_A - \mathbf{t}_B) = r_a + d_b, \tag{3.1}$$

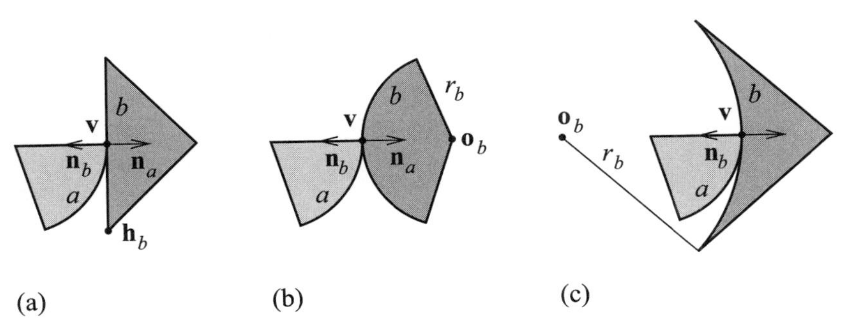

Figure 3.3
Boundary conditions: (a) line, (b–c) circle.

where $d_b = \mathbf{n}_b^B \cdot \mathbf{h}_b^B$ replaces $(R_B\mathbf{n}_b^B) \cdot (R_B\mathbf{h}_b^B)$ using the invariance of inner products under rotation.

Figure 3.2b–c shows the tangency condition for a circle, a, with radius r_a and center $\mathbf{o}_a$, and a circle, b, with radius r_b and center $\mathbf{o}_b$. When the circles touch on the outside (figure 3.2b), the distance between the centers equals the sum of the radii: $\|\mathbf{o}_b - \mathbf{o}_a\| = r_a + r_b$ with $\|\cdot\|$ denoting the vector norm $\|\mathbf{v}\| = (v_x^2 + v_y^2)^{1/2}$. When the outside of a touches the inside of b (figure 3.2c), the distance between the centers equals the difference of the radii, $\|\mathbf{o}_b - \mathbf{o}_a\| = r_b - r_a$, and $r_b > r_a$. We express these equations in part coordinates by substituting

$$\mathbf{o}_a = R_A\mathbf{o}_a^A + \mathbf{t}_A$$

$$\mathbf{o}_b = R_B\mathbf{o}_b^B + \mathbf{t}_B$$

to obtain

$$\|R_B\mathbf{o}_b^B + \mathbf{t}_B - R_A\mathbf{o}_a^A - \mathbf{t}_A\| = r_b \pm r_a. \tag{3.2}$$

3.1.2 Boundary Equations

Figure 3.3a shows the boundary conditions for a vertex, $\mathbf{v}$, and a line, b. The vertex is on the line and the outward normals, $\mathbf{n}_a$ and $\mathbf{n}_b$, are opposite. The equations are $\mathbf{n}_b \cdot (\mathbf{v} - \mathbf{h}_b) = 0$ and $\mathbf{n}_b = -\mathbf{n}_a$. Substituting part coordinates yields

$$(R_B\mathbf{n}_b^B) \cdot (\mathbf{t}_A + R_A\mathbf{v}^A - \mathbf{t}_B) = d_b \tag{3.3}$$

$$R_B\mathbf{n}_b^B = -R_A\mathbf{n}_a^A, \tag{3.4}$$

with $d_b = \mathbf{n}_b^B \cdot \mathbf{h}_b^B$.

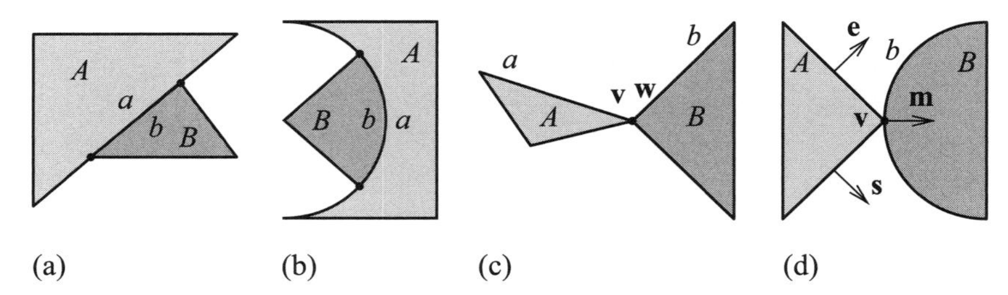

Figure 3.4
Subsumed contacts: (a) line-line, (b) identical circles, (c) vertex-vertex, (d) segment-vertex.

Figure 3.3b shows the boundary conditions for a vertex, $\mathbf{v}$, and a circle, b. The vertex is on the line and the outward normals are opposite. The equations are $\mathbf{o}_b \pm r_b \mathbf{n}_b = \mathbf{v}$, using the circle property that $\mathbf{v} - \mathbf{o}_b$ is colinear with $\mathbf{n}_b$, and $\mathbf{n}_b = -\mathbf{n}_a$. Substituting the second equation into the first and introducing part coordinates yields

$$\mathbf{t}_A + R_A \mathbf{v}^A \pm r_b R_A \mathbf{n}_a^A = \mathbf{t}_B + R_B \mathbf{o}_b^B$$

and the linearity of rotation yields

$$\mathbf{t}_A + R_A \mathbf{e} = \mathbf{t}_B + R_B \mathbf{o}_b^B, \tag{3.5}$$

with $\mathbf{e} = \mathbf{v}^A \pm r_b \mathbf{n}_a^A$.

3.1.3 Subsumed Contacts

We need not compute line-line contacts because they are also line-vertex contacts. Either both vertices of the shorter segment are in contact with the longer segment or one vertex of each segment is in contact with the other segment (figure 3.4a). For the same reason, we need not compute contacts for identical circles (figure 3.4b) because they are also circle-vertex contacts. The analysis in later sections implicitly ignores this case. We need not compute vertex-vertex contacts because they are also segment-vertex contacts. When vertex $\mathbf{v}$ of a is in contact with vertex $\mathbf{w}$ of b, $\mathbf{v}$ is in contact with b and a is in contact with $\mathbf{w}$ (figure 3.4c).

We rewrite a segment-vertex contact as a contact between the segment and an artificial circle with radius zero. A contact between vertex $\mathbf{v}$ and segment b occurs when $\mathbf{v}$ lies on b and the b inward normal, $\mathbf{m}$, is between the outward normals, $\mathbf{s}$ and $\mathbf{e}$, of the two incident segments (figure 3.4d). This is mathematically equivalent to contact between b and a circle segment with radius zero, center and vertices $\mathbf{v}$, and vertex normals $\mathbf{s}$ and $\mathbf{e}$.

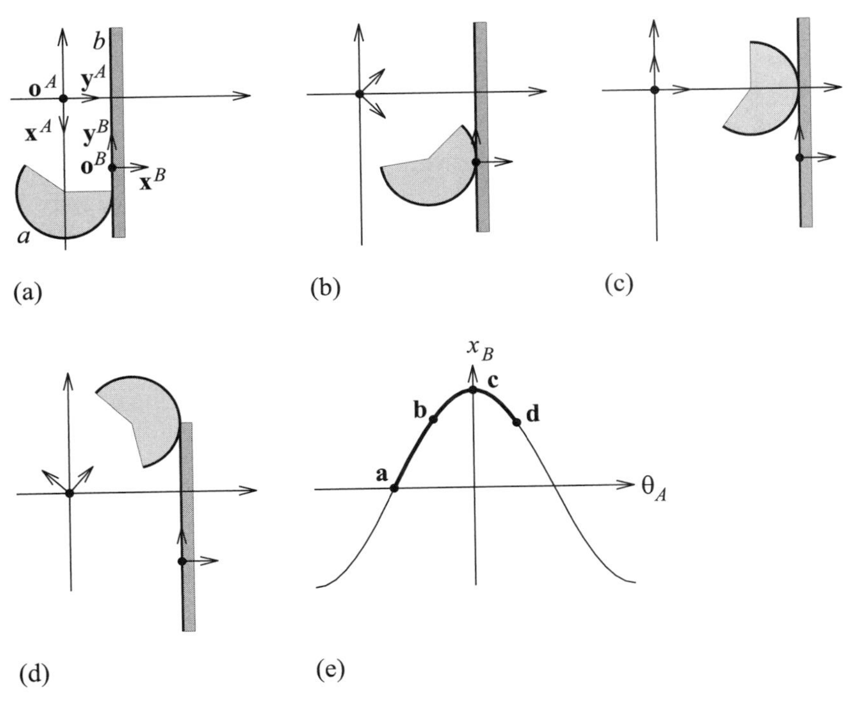

Figure 3.5
Fixed-axis pair: (a–d) contact configurations **a–d**, (e) contact curve (thick) and tangency curve (thin).

3.2 Fixed-Axis Planar Pairs

A fixed-axes pair has a single tangency equation whose variables are the two degrees of freedom, while the other part coordinates are constant. The solution set is a curve in the pair's configuration space, called a tangency curve. The solution set of the boundary equations is a subcurve of the tangency curve, called a contact curve. Figure 3.5a–d shows contact configurations **a–d** of a rotating circle segment, *a*, and a translating line segment, *b*. Figure 3.5e shows the tangency and contact curves. The contact point coincides with an *a* vertex in configuration **a**, is interior to both segments in configurations **b** and **c**, and coincides with a *b* vertex in configuration **d**.

3.2.1 Tangency Curves

We solve a tangency equation by writing one variable as a closed-form expression of the other variable on a closed interval. Although the example given here yields one interval, other equations yield several intervals. We saw in section 2.3 that there are three types of fixed-axis pairs: rotation-rotation, rotation-x translation, and x translation-y translation. These types combine with the two tangency equations to generate eight cases: rotating circle-translating line, rotating circle-rotating circle, and so on. In this section we solve the five cases involving rotation. We solve the three translation cases at the end of section 3.3 since their main use is in general planar pairs.

Rotating Circle-Translating Line The degrees of freedom are θ_A, x_B and the configuration variables x_A, y_A, y_B, θ_B are constant. Since θ_B is constant, $\mathbf{n} = R_B\mathbf{n}_b^B$ is constant. Also $\mathbf{t}_A = (x_A, y_A)$ is constant. Substituting $\mathbf{n}$ in equation 3.1, expanding the parentheses, and rearranging yields

$$\mathbf{n} \cdot \mathbf{t}_B = \mathbf{n} \cdot (R_A\mathbf{o}_a^A + \mathbf{t}_A) - r_a - d_b.$$

Expanding the left side and rearranging yields

$$n_x x_B = \mathbf{n} \cdot (R_A\mathbf{o}_a^A + \mathbf{t}_A) - r_a - d_b - n_y y_B.$$

We expand $\mathbf{t}_A$ and employ the identity

$$\mathbf{n} \cdot (R_A\mathbf{v}) = (\mathbf{v} \times \mathbf{n}) \sin\theta_A + (\mathbf{v} \cdot \mathbf{n}) \cos\theta_A$$

to obtain

$$n_x x_B = (\mathbf{o}_a^A \times \mathbf{n}) \sin\theta_A + (\mathbf{o}_a^A \cdot \mathbf{n}) \cos\theta_A - r_a - d_b - n_y y_B + n_x x_A + n_y y_A.$$

Dividing by n_x gives the closed-form solution for x_B in terms of θ_A and the four constant configuration variables: $x_B = k_1 \sin\theta_A + k_2 \cos\theta_A + k_3$.

Figure 3.6 shows a typical pair. The constant configuration variables are $x_A = 0$, $y_A = 0$, $y_B = 0$, and $\theta_B = 0$. The curve parameters are $\mathbf{o}_a^A = (2, 0)$, $r_a = 1$, $\mathbf{h}_b^B = (0, 0)$, and $\mathbf{n}_b^B = (1, 0)$. The curve, $x_B = 2\cos\theta_A - 1$, is a sinusoid.

A degeneracy occurs when the line b is parallel to the $\mathbf{x_B}$ axis, so $n_x = 0$. The tangency curve is two vertical lines where the right side of the tangency equation equals zero. Figure 3.7 shows a degenerate tangency curve. The only change from the previous example is that b is horizontal with $\mathbf{h}_b^B = (0, -2)$ and $\mathbf{n}_b^B = (0, 1)$. The right-side equation is $2\cos\theta_A = -1$ with zeros at $-150°$ and $-30°$.

Translating Circle-Rotating Line The degrees of freedom are x_A, θ_B and the configuration variables y_A, θ_A, x_B, y_B are constant. Since θ_A is constant,

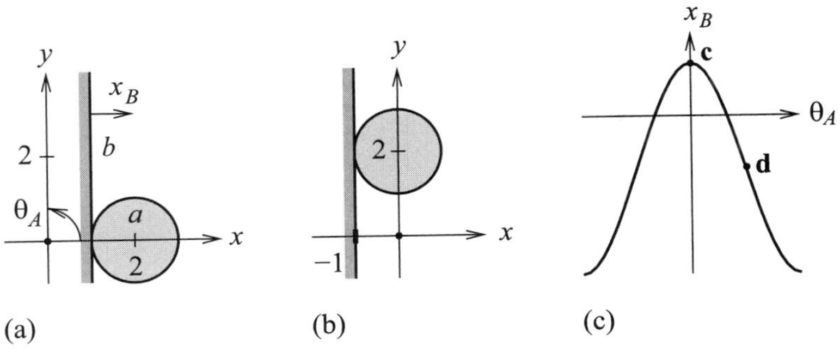

Figure 3.6
Rotating circle-translating line: (a) configuration $\mathbf{c} = (0°, 1)$, (b) configuration $\mathbf{d} = (90°, -1)$, (c) tangency curve.

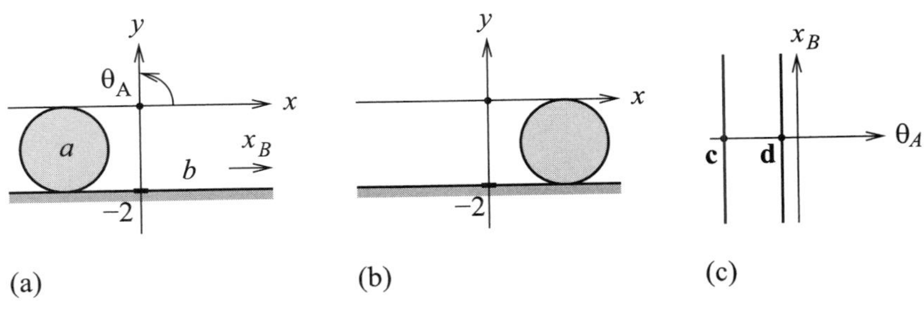

Figure 3.7
Degenerate case: (a) configuration $\mathbf{c} = (-150°, 0)$, (b) configuration $\mathbf{d} = (-30°, 0)$, (c) tangency curve.

$\mathbf{o} = R_A \mathbf{o}_a^A$ is constant. Also $\mathbf{t}_B = (x_B, y_B)$ is constant. Substituting $\mathbf{n} = R_B \mathbf{n}_b^B$ and $\mathbf{o}$ in equation 3.1, expanding the parentheses, and rearranging yields

$$\mathbf{n} \cdot \mathbf{t}_A = \mathbf{n} \cdot (\mathbf{t}_B - \mathbf{o}) + r_a + d_b.$$

Expanding the left side and rearranging yields

$$n_x x_A = \mathbf{n} \cdot (\mathbf{t}_B - \mathbf{o}) + r_a + d_b - n_y y_A.$$

Expanding $\mathbf{n}$ yields the closed-form solution for x_A in terms of θ_B.

$$x_A = \frac{[\mathbf{n}_b^B \times (\mathbf{t}_B - \mathbf{o})] \sin\theta_B + [\mathbf{n}_b^B \cdot (\mathbf{t}_B - \mathbf{o})] \cos\theta_B + r_a + d_b - n_y y_A}{n_x^B \cos\theta_B - n_y^B \sin\theta_B}.$$

Figure 3.8 shows a typical pair. The constant configuration variables are $y_A = 0$, $\theta_A = 0$, $x_B = 0$, and $y_B = 0$. The curve parameters are $\mathbf{o}_a^A = (2, 0)$,

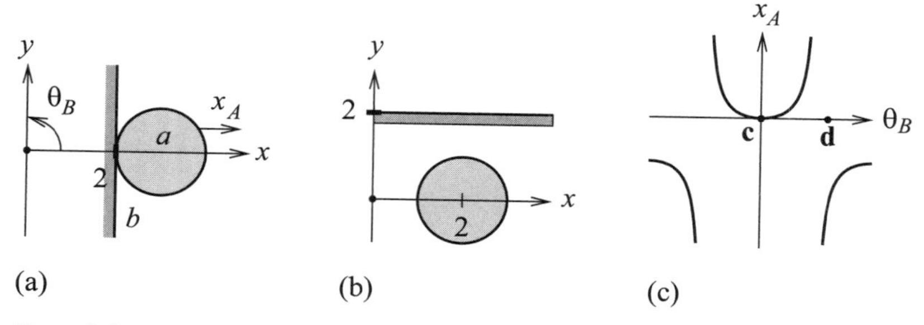

Figure 3.8
Translating circle-rotating line: (a) configuration $\mathbf{c} = (0, 0)$, (b) configuration $\mathbf{d} = (90°, 0)$, (c) tangency curve.

$r_a = 1$, $\mathbf{h}_b^B = (2, 0)$, and $\mathbf{n}_b^B = (1, 0)$. The solution is $x_A = (3 - 2\cos\theta_B)/\cos\theta_B$. The tangency curve has vertical asymptotes at $\theta_B = \pm 90°$ where the denominator equals zero. There is no tangency for any x_A value because a translates parallel to b (figure 3.8b).

Rotating Circle-Translating Circle The degrees of freedom are θ_A, x_B and the configuration variables x_A, y_A, y_B, θ_B are constant. Hence, $\mathbf{e} = R_B\mathbf{o}_b^B - \mathbf{t}_A$ is constant. Define $\mathbf{f} = \mathbf{e} - R_A\mathbf{o}_a^A$. We employ the identity $\|\mathbf{v}\|^2 = \mathbf{v}^2$ where $\mathbf{v}^2$ denotes $\mathbf{v}\cdot\mathbf{v}$. Substituting $\mathbf{f}$ in equation 3.2 and squaring both sides of the equation yields

$$(\mathbf{t}_B + \mathbf{f})^2 = (r_b \pm r_a)^2.$$

Expanding the inner product yields

$$\mathbf{t}_B^2 + 2\mathbf{f}\cdot\mathbf{t}_B + \mathbf{f}^2 - (r_b \pm r_a)^2 = 0.$$

Expanding the three inner products yields

$$\mathbf{t}_B^2 = x_B^2 + y_B^2$$

$$\mathbf{f}\cdot\mathbf{t}_B = x_B(e_x + o_{ay}^A \sin\theta_A - o_{ax}^A \cos\theta_A) + y_B(e_y - o_{ax}^A \sin\theta_A - o_{ay}^A \cos\theta_A)$$

$$\mathbf{f}^2 = \mathbf{e}^2 + (\mathbf{o}_a^A)^2 - 2(\mathbf{o}_a^A \times \mathbf{e})\sin\theta_A - 2(\mathbf{o}_a^A\cdot\mathbf{e})\cos\theta_A.$$

Substituting the expansions into the tangency equation yields

$$x_B^2 + 2(e_x + o_{ay}^A \sin\theta_A - o_{ax}^A \cos\theta_A)x_B - 2(o_{ax}^A y_B + \mathbf{o}_a^A \times \mathbf{e})\sin\theta_A$$
$$- 2(o_{ay}^A y_B + \mathbf{o}_a^A\cdot\mathbf{e})\cos\theta_A + \mathbf{e}^2 + (\mathbf{o}_a^A)^2 + 2e_y y_B - (r_b \pm ra)^2 = 0.$$

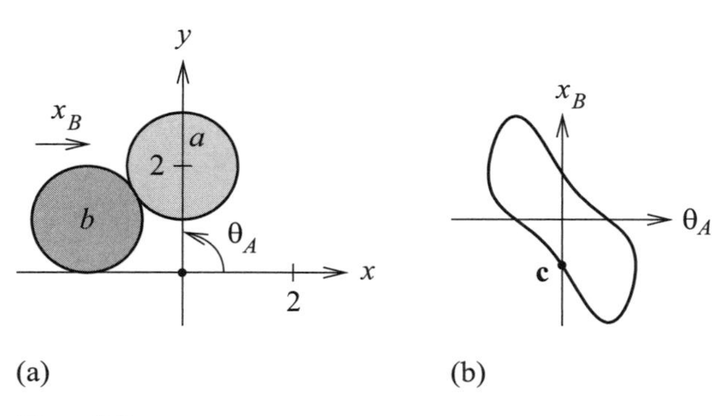

Figure 3.9
Rotating circle-translating circle: (a) configuration $\mathbf{c} = (0°, -1.73)$, (b) tangency curves.

The tangency equation is a quadratic in x_B whose coefficients are linear in $\sin\theta_A$ and $\cos\theta_A$. The domain of the tangency curves is the θ_A intervals where the discriminant is positive. The discriminant is quadratic in $\sin\theta_A$ and $\cos\theta_A$; hence it has at most four real zeroes that define at most two domain intervals, each of which yields two tangency curves.

Figure 3.9 shows a typical pair. The constant configuration variables are $x_A = 0$, $y_A = 0$, $y_B = 0$, and $\theta_B = 0$. The curve parameters are $\mathbf{o}_a^A = (0, 2)$, $r_a = 1$, $\mathbf{o}_b^B = (0, 1)$, and $r_b = 1$. The tangency equation is $x_B^2 + (4\sin\theta_A)x_B - 4\cos\theta_A + 1 = 0$. The discriminant, $16\sin^2\theta_A + 16\cos\theta_A - 4$, is zero at $\theta_A = \pm 120°$, so the domain of the two tangency curves is $[-120°, 120°]$.

Rotating Circle-Rotating Line The degrees of freedom are θ_A, θ_B, with x_A, y_A, x_B, y_B constant. Substituting $\mathbf{e} = R_A\mathbf{o}_a^A + \mathbf{t}_A - \mathbf{t}_B$ and $k_3 = r_a + d_b$ in equation 3.1 and expanding R_B yields

$$(\mathbf{n}_b^B \times \mathbf{e})\sin\theta_B + (\mathbf{n}_b^B \cdot \mathbf{e})\cos\theta_B = k_3.$$

Expanding $\mathbf{e}$ and R_A yields

$$k_1\sin\theta_B + k_2\cos\theta_B = k_3, \tag{3.6}$$

with

$$k_1 = (\mathbf{n}_b^B \cdot \mathbf{o}_a^A)\sin\theta_A + (\mathbf{n}_b^B \times \mathbf{o}_a^A)\cos\theta_A + \mathbf{n}_b^B \times (\mathbf{t}_A - \mathbf{t}_B)$$

$$k_2 = (\mathbf{n}_b^B \cdot \mathbf{o}_a^A)\cos\theta_A - (\mathbf{n}_b^B \times \mathbf{o}_a^A)\sin\theta_A + \mathbf{n}_b^B \cdot (\mathbf{t}_A - \mathbf{t}_B).$$

Let $k_4 = (k_1^2 + k_2^2)^{1/2}$ and define α by $\sin\alpha = k_2/k_4$ and $\cos\alpha = k_1/k_4$. Dividing the tangency equation by k_4 yields

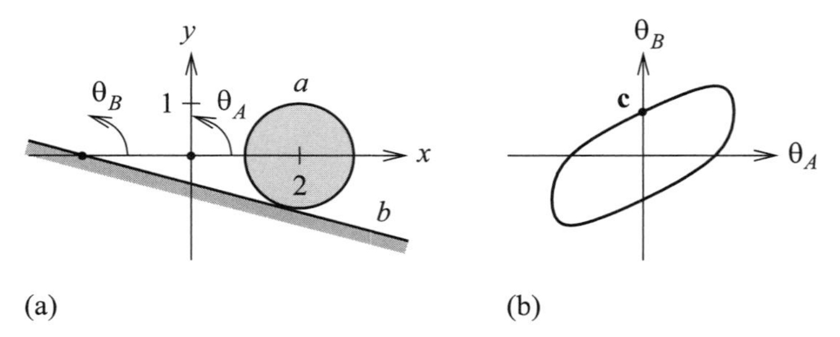

Figure 3.10
Rotating circle-rotating line: (a) configuration $\mathbf{c} = (0°, 75.52°)$, (b) tangency curves.

$$\cos\alpha \sin\theta_B + \sin\alpha \cos\theta_B = k_3/k_4$$

or $\sin(\theta_B + \alpha) = k_3/k_4$. The domain of the tangency curve is the θ_A intervals where $|k_3/k_4| \leq 1$ or $k_3^2 \leq k_1^2 + k_2^2$. The two solutions are $\arcsin(k_3/k_4) - \alpha$ and $180° - \arcsin(k_3/k_4) - \alpha$.

Figure 3.10 shows a typical pair. The constant configuration variables are $x_A = 0$, $y_A = 0$, $x_B = -2$ and $y_B = 0$. The curve parameters are $\mathbf{o}_a^A = (2,0)$, $r_a = 1$, $\mathbf{h}_b^B = (0,0)$, and $\mathbf{n}_b^B = (1,0)$. The tangency equation is

$$2\sin\theta_A \sin\theta_B + 2(\cos\theta_A + 1)\cos\theta_B = 1.$$

The domain inequality, $8(1 + \cos\theta_A) \geq 1$, holds on the interval $[-151°, 151°]$.

Rotating Circle-Rotating Circle The degrees of freedom are θ_A, θ_B, with x_A, y_A, x_B, y_B constant. Substituting $\mathbf{t} = \mathbf{t}_B - \mathbf{t}_A$, $\mathbf{p} = R_A\mathbf{o}_a^A$, and $\mathbf{q} = R_B\mathbf{o}_b^B$ in equation 3.2 yields

$$(\mathbf{h} + \mathbf{t} - \mathbf{p})^2 = (r_b \pm r_a)^2.$$

Expansion yields

$$(\mathbf{o}_b^B)^2 + \mathbf{t}^2 + (\mathbf{o}_a^A)^2 + 2\mathbf{q}\cdot\mathbf{t} - 2\mathbf{p}\cdot\mathbf{t} - 2\mathbf{q}\cdot\mathbf{p} = (r_b \pm r_a)^2$$

and rearrangement yields

$$\mathbf{q}\cdot(\mathbf{p} - \mathbf{t}) + \mathbf{p}\cdot\mathbf{t} = k_4,$$

with

$$k_4 = 0.5[(\mathbf{o}_b^B)^2 + \mathbf{t}^2 + (\mathbf{o}_a^A)^2 - (r_b \pm r_a)^2].$$

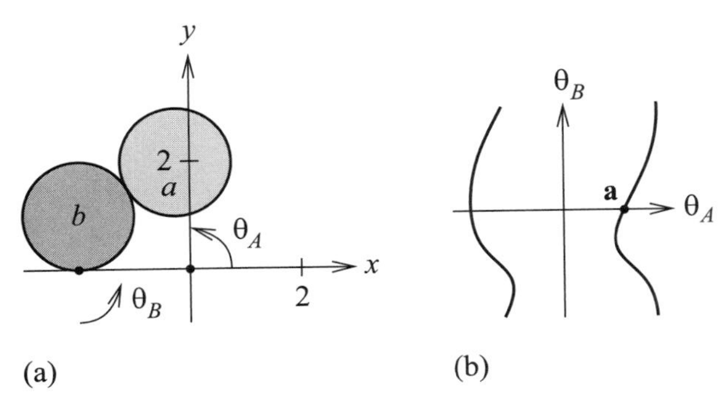

Figure 3.11
Rotating circle-rotating circle: (a) configuration $\mathbf{c} = (97.4°, 0°)$, (b) tangency curves.

Expanding $\mathbf{q}$ yields

$$[\mathbf{o}_b^B \times (\mathbf{p} - \mathbf{t})] \sin \theta_B + [\mathbf{o}_b^B \cdot (\mathbf{p} - \mathbf{t})] \cos \theta_B + \mathbf{p} \cdot \mathbf{t} = k_4$$

then expanding $\mathbf{p}$ yields equation 3.6 with

$$k_1 = (\mathbf{o}_a^A \cdot \mathbf{o}_b^B) \sin \theta_A - (\mathbf{o}_a^A \times \mathbf{o}_b^B) \cos \theta_A - \mathbf{o}_b^B \times \mathbf{t}$$

$$k_2 = (\mathbf{o}_a^A \times \mathbf{o}_b^B) \sin \theta_A + (\mathbf{o}_a^A \cdot \mathbf{o}_b^B) \cos \theta_A - \mathbf{o}_b^B \cdot \mathbf{t}$$

$$k_3 = k_4 - (\mathbf{o}_a^A \times \mathbf{t}) \sin \theta_A - (\mathbf{o}_a^A \cdot \mathbf{t}) \cos \theta_A.$$

Figure 3.11 shows a typical pair. The constant configuration variables are $x_A = 0$, $y_A = 0$, $x_B = -2$, and $y_B = 0$. The curve parameters are $\mathbf{o}_a^A = (2, 0)$, $r_a = 1$, $\mathbf{o}_b^B = (0, 1)$, and $r_b = 1$. The tangency equation is

$$-2(1 + \cos \theta_A) \sin \theta_B + 2 \sin \theta_A \cos \theta_B = 2.5 + 4 \cos \theta_A.$$

The domain inequality, $16 \cos^2 \theta_A + 12 \cos \theta_A \geq 1.75$, holds on the intervals $[-151°, -83°]$ and $[83°, 151°]$.

A degeneracy occurs when $\mathbf{o}_b^B = (0, 0)$, since $k_1 = 0$, $k_2 = 0$ for all θ_A values. The tangency curve is the solution of $k_3 = 0$, which is two vertical lines. When $\mathbf{o}_a^A = (0, 0)$ as well, there are no tangency curves.

3.2.2 Contact Curves

We compute contact curves by solving the boundary equations and identifying the solutions that are boundary configurations. The solutions are the configurations where a vertex of one segment lies on a circle or line of the other segment. A solution is a boundary configuration when the vertex lies

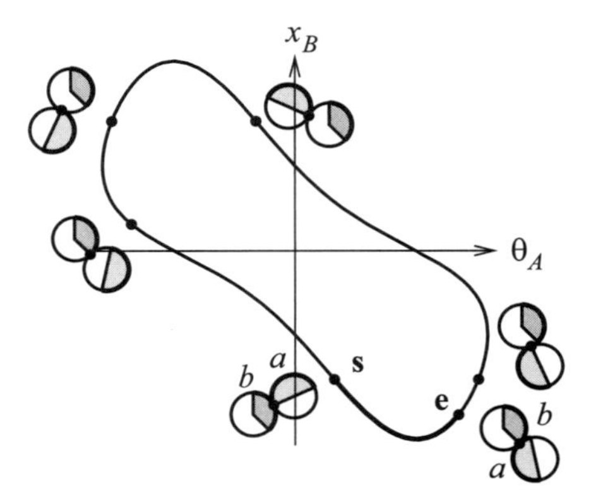

Figure 3.12
Contact curve (thick), tangency curve (thin), and boundary equation solutions (dots plus feature snapshots).

on the segment. For a line segment, the vertex coordinates are between those of the segment's vertices. For a circle segment, the ray from the circle's center to the vertex is between the rays from the circle's center to its tail and head. We split the tangency curve at the boundary configurations. The contact curves are the subcurves whose midpoints are contact configurations.

Figure 3.12 illustrates this for on a rotating circle segment, *a*, and a translating circle segment, *b*. There are six boundary equation solutions, labeled with dots, and two boundary configurations: **s** where an *a* vertex is on *b* and **e** where a *b* vertex is on *a*. The **se** subcurve is the contact curve.

Solving Boundary Equations There is one set of boundary equations for a vertex-line contact and another for a vertex-circle contact. The three types of fixed-axis pairs yield three cases for each equation: rotating vertex-translating segment, translating vertex-rotating segment, and rotating vertex-rotating segment.

The vertex-line equations are 3.3 and 3.4. When A translates and B rotates, $\mathbf{n}_a$ is constant. Equation 3.4 implies that $\mathbf{n}_a \times R_B \mathbf{n}_b^B = 0$, which has the form $k_1 \sin\theta_B + k_2 \cos\theta_B = 0$ with k_1 and k_2 constant; hence it is solved like equation 3.6. Once θ_B is known, equation 3.3 is linear in the second unknown, x_A. When A rotates and B translates, $\mathbf{n}_b$ is constant, equation 3.4 yields θ_A, and equation 3.3 yields x_B. When both parts rotate, substituting the second equation into the first and using the invariance of inner products under rotation yields

$$(R_A\mathbf{n}_a^A) \cdot (\mathbf{t}_A - \mathbf{t}_B) = d_b - \mathbf{n}_a^A \cdot \mathbf{v}^A.$$

This equation has the form $k_1 \sin\theta_A + k_2 \cos\theta_A = k_3$ with k_1, k_2, k_3 constant, which is solved as earlier. Given θ_A, we compute $\mathbf{n}_a$ and obtain θ_B from the second equation as before.

The vertex-circle equation is equation 3.5. When A and B rotate, the left side of the equation defines a circle parameterized by θ_A with center $\mathbf{t}_A$ and radius $\|\mathbf{e}\|$, while the right side defines a circle parameterized by θ_B. Intersecting these circles (a standard computation) yields the solutions to the boundary equations. When A rotates and B translates, the equations are

$$x_A + e_x \cos\theta_A - e_y \sin\theta_A = x_B + o_x$$

$$y_A + e_x \sin\theta_A + e_y \cos\theta_A = y_B + o_y,$$

with $\mathbf{o} = R_B\mathbf{o}_b^B$ constant. The second equation is solved for θ_A, then the first is solved for x_B. Likewise when A translates and B rotates.

Figure 3.12 shows typical rotating arc-translating arc contact curves. The constant configuration variables are $x_A = 0$, $y_A = 0$, $y_B = 0$, and $\theta_B = 0$. The curve parameters are $\mathbf{o}_a^A = (0, 2)$, $r_a = 1$, $\mathbf{t}_a^A = (1, 2)$, $\mathbf{h}_a^A = (-1, 2)$, $\mathbf{o}_b^B = (0, 1)$, $r_b = 1$, $\mathbf{t}_b^B = (0.71, 0.29)$, and $\mathbf{h}_b^B = (0, 2)$. The equations for $\mathbf{v} = \mathbf{h}_a$ are

$$-2\cos\theta_A - 2\sin\theta_A = x_B$$

$$-2\sin\theta_A + 2\cos\theta_A = 1.$$

The first solution is the boundary configuration $\mathbf{s} = (24.3°, -2.64)$ and the second is $\mathbf{c} = (-114°, 2.64)$ where $\mathbf{h}_a$ is only on the b circle.

3.3 General Planar Pairs

We analyze general planar pairs in the (x_A, y_A, θ_A) configuration space of part A relative to part B (figure 2.16). The segment contact equations define contact surfaces in the configuration space. The cross-section at $\theta_A = \theta_0$ of the a/b contact surface consists of the contact configurations where a translates at orientation θ_0. It is a contact curve in the (x_A, y_A) configuration space of a rotated version of a that translates relative to b. The curve is a circle segment when both a and b are circle segments and is a line segment otherwise. We exploit this special structure to obtain simple closed-form expressions for contact surfaces.

We derive closed-form expressions for the three types of contact: circle segment-line segment, line segment-circle segment, and circle segment-circle segment. A contact surface is represented as a contact curve

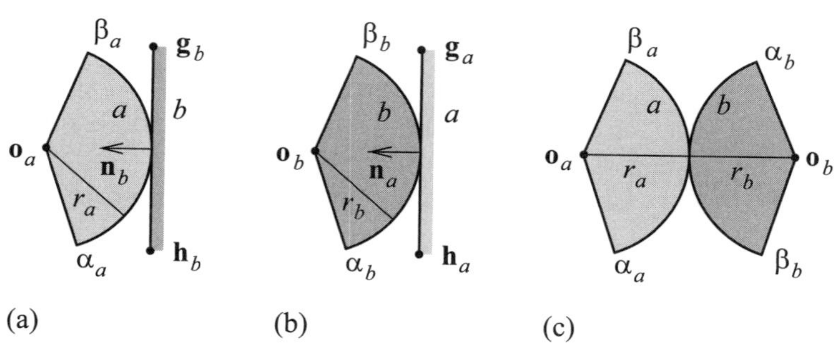

Figure 3.13
General segment contacts: (a) moving circle-fixed line, (b) moving line-fixed circle, (c) moving circle-fixed circle.

parameterized by θ_A and defined over an interval of θ_A values. The derivation of the parameterized contact curve follows the strategy of the previous section: compute the tangency curve, compute the boundary points, and identify the portion of the tangency curve that is the contact curve. The main difference is that the rotated a segment contains θ_A as a parameter.

Moving Circle Segment-Fixed Line Segment Figure 3.13a shows the contact conditions for a circle segment, a, with radius r_a and center $\mathbf{o}_a$, and a line segment, b, with point $\mathbf{h}_b$ and unit normal $\mathbf{n}_b$. Tangency occurs when the distance from $\mathbf{o}_a$ to the line equals r_a. The tangency equation is

$$\mathbf{n}_b(R_A\mathbf{o}_a^A + \mathbf{t}_A) = r_a + d_b,$$

using equation 3.1 with $\mathbf{t}_B = (0, 0)$ and $\theta_B = 0$. The tangency curve,

$$n_{bx}x_A + n_{by}y_A + \mathbf{n}_b \cdot R_A\mathbf{o}_a^A - r_a - d_b = 0, \tag{3.7}$$

is a line parallel to b whose distance from the origin varies with θ_A.

Contact occurs when the point of tangency lies on a: the b inward normal angle, $\nu = \arctan(-n_{by}, -n_{bx})$, is in the interval of a outward normal angles, $[\alpha_a, \beta_a] = [\alpha_a^A + \theta_A, \beta_a^A + \theta_A]$, so $\theta_A \in [\nu - \beta_a^A, \nu - \alpha_a^A]$. The boundary configurations are where the point of tangency coincides with the vertices of b. The equations are $\mathbf{o}_a - r_a\mathbf{n}_b = \mathbf{g}_b$ and $\mathbf{o}_a - r_a\mathbf{n}_b = \mathbf{h}_b$, as in section 3.2.2. The boundary configurations are $\mathbf{s} = \mathbf{g}_b + \mathbf{q}$ and $\mathbf{e} = \mathbf{h}_b + \mathbf{q}$ with $\mathbf{q} = r_a\mathbf{n}_b - R_A\mathbf{o}_a^A$. Point $\mathbf{q}$ lies on a circle of radius $\|\mathbf{o}_a^A\|$ centered at $r_a\mathbf{n}_b$. As θ varies, $\mathbf{s}$ and $\mathbf{e}$ rotate on translated copies of this circle. The contact curve, $\mathbf{se}$, is b translated by $\mathbf{q}$.

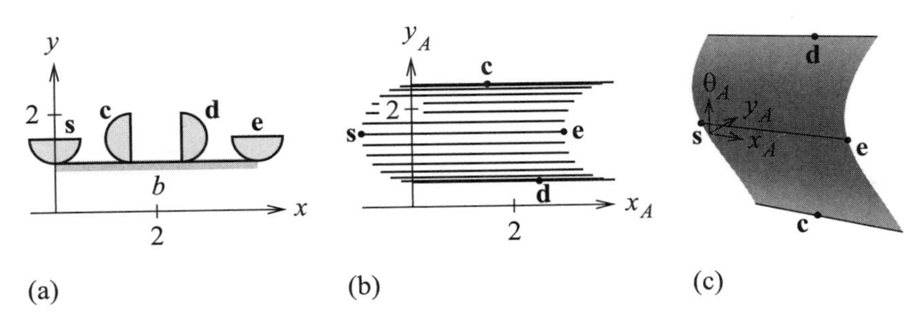

Figure 3.14
Moving circle segment-fixed line segment: (a) contact configurations, (b) contact curves, (c) contact surface.

Figure 3.14a shows a typical case for boundary configurations. Segment a has center $\mathbf{o}_a^A = (1, 0)$, radius $r_a = 0.5$, start angle $\alpha_a^A = -180°$, and end angle $\beta_a^A = 0°$. Segment b has vertices $\mathbf{g}_b = (0, 1)$ and $\mathbf{h}_b = (4, 1)$, outward normal $\mathbf{n}_b = (0, 1)$, and inward normal angle $\nu = -90°$. The tangency curve is $y_A + \sin\theta_A - 1.5 = 0$. Contact occurs for $\theta_A \in [-90°, 90°]$ with boundary configurations $\mathbf{s} = (0, 1) + R_A(-1, -0.5)$ and $\mathbf{e} = (0, 1) + R_A(1, -0.5)$. The displayed configurations are $\mathbf{s} = (-1, 1.5, 0°)$, $\mathbf{e} = (3, 1.5, 0°)$, $\mathbf{c} = (1.5, 2.5, -90°)$, and $\mathbf{d} = (2.5, 0.5, 90°)$. Figure 3.14b shows the parametric contact curve at selected θ_A values and figure 3.14c shows the contact surface.

Moving Line Segment-Fixed Circle Segment Figure 3.13b shows the contact conditions for a line segment, a, with point $\mathbf{h}_a$ and unit normal $\mathbf{n}_a$, and a circle segment, b, with radius r_b and center $\mathbf{o}_b$. Tangency occurs when the distance from $\mathbf{o}_b$ to the line equals r_b. The tangency equation is

$$\mathbf{n}_a \cdot (\mathbf{o}_b - \mathbf{h}_a) = r_b,$$

which yields

$$(n_{ax}^A \cos\theta_a - n_{ay}^A \sin\theta_a)(x_A - o_{bx}) + (n_{ax}^A \sin\theta_a + n_{ay}^A \cos\theta_a)(y_A - o_{by})$$
$$+ r_b - d_a = 0,$$

with $d_a = \mathbf{n}_a^A \cdot \mathbf{h}_a^A$. The tangency curve is a line that forms an angle of θ_A with a and whose distance from the origin is $|r_b - d_a|$. Its inward normal angle is $\nu + \theta_A$ with $\nu = \arctan(-n_{ay}^A, -n_{ax}^A)$ the inward normal angle of a.

Contact occurs when the point of tangency lies on b: $\nu + \theta_A$ is in the interval of b outward normal angles, $[\alpha_b, \beta_b]$, so $\theta_A \in [\alpha_b - \nu, \beta_b - \nu]$. The boundary configurations are where the point of tangency coincides with

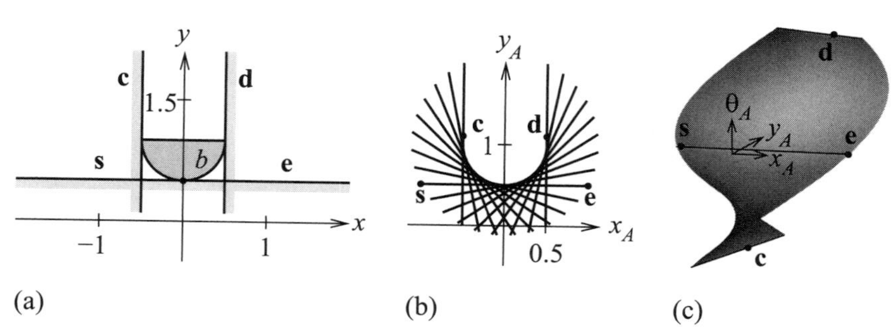

Figure 3.15
Moving line segment-fixed circle segment: (a) contact configurations, (b) contact curves, (c) contact surface.

the vertices of a. The equations are $\mathbf{o}_b - r_b\mathbf{n}_a = \mathbf{g}_a$ and $\mathbf{o}_b - r_b\mathbf{n}_a = \mathbf{h}_a$. The boundary configurations,

$$\mathbf{s} = \mathbf{o}_b + R_A(\mathbf{g}_a^A - r_b\mathbf{n}_a^A)$$

$$\mathbf{e} = \mathbf{o}_b + R_A(\mathbf{h}_a^A - r_b\mathbf{n}_a^A),$$

rotate on circles centered at $\mathbf{o}_b$. The contact curve, **se**, is a rotated by θ_A around $r_b\mathbf{n}_a^A$ then translated by $\mathbf{o}_b - r_b\mathbf{n}_a^A$.

Figure 3.15a shows a typical case. Segment a has vertices $\mathbf{g}_a^A = (-1, 0)$ and $\mathbf{h}_a^A = (1, 0)$, outward normal $\mathbf{n}_a^A = (0, 1)$, inward normal angle $\nu = -90°$, and $d_a = 0$. Segment b has a center $\mathbf{o}_b = (0, 1)$, radius $r_b = 0.5$, start angle $\alpha_b = -180°$, and end angle $\beta_b = 0°$. The tangency curve is $x_A \sin\theta_A + (1 - y_A)\cos\theta_A = 0.5$. Contact occurs for $\theta_A \in [-90°, 90°]$ with boundary configurations $\mathbf{s} = (0, 1.5) - (\cos\theta_A, \sin\theta_A)$ and $\mathbf{e} = (4, 1.5) - (\cos\theta_A, \sin\theta_A)$. The displayed configurations are $\mathbf{s} = (-1, 1.5, 0°)$, $\mathbf{e} = (3, 1.5, 0°)$, $\mathbf{c} = (1.5, 2.5, -90°)$, and $\mathbf{d} = (2.5, 0.5, 90°)$. Figure 3.15b–c shows the parametric contact curve and the contact surface.

Moving Circle Segment-Fixed Circle Segment Figure 3.13c shows the contact conditions for a circle segment, a, with radius r_a and center $\mathbf{o}_a$, and a circle segment, b, with radius r_b and center $\mathbf{o}_b$. The distance between the centers equals the sum or difference of the radii when the circles touch on the outside or on the inside. The tangency equation is

$$\|\mathbf{t}_A + R_A\mathbf{o}_a^A - \mathbf{o}_b\| = r \quad \text{with } r = |r_b \pm r_a|, \tag{3.8}$$

using equation 3.1 with $\mathbf{t}_B = (0, 0)$ and $\theta_B = 0$. The tangency curve is a circle of radius r centered at $\mathbf{o} = \mathbf{o}_b - R_A\mathbf{o}_a^A$. As θ increases, it rotates on a circle

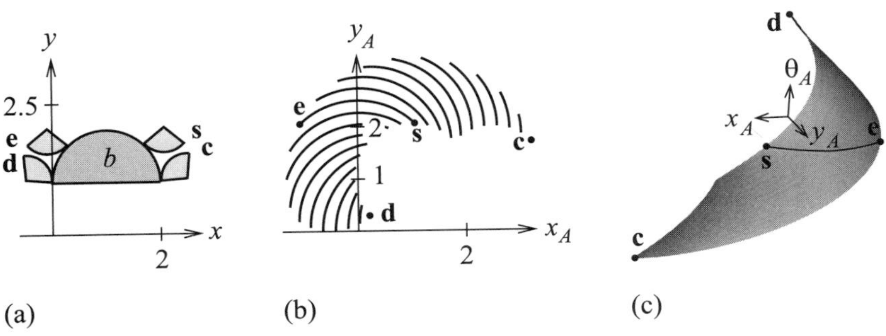

Figure 3.16
Moving circle segment-fixed circle segment: (a) contact configurations, (b) contact curves, (c) contact surface.

of radius $\|\mathbf{o}_a^A\|$ centered at $\mathbf{o}_b$. To derive the contact curve, we parameterize the tangency curve as $\mathbf{q}(\gamma) = \mathbf{o} + r(\cos\gamma, \sin\gamma)$.

We derive the contact curve for two circle segments whose combined angular extent, $\beta_a^A - \alpha_a^A + \beta_b - \alpha_b$, is at most 360°. Otherwise, we split the segments in half and analyze the four subsegment contacts. In contact configuration $\mathbf{t}_A = \mathbf{q}(\gamma)$,

$$\begin{aligned}\mathbf{o}_a - \mathbf{o}_b &= \mathbf{q}(\gamma) + R_A\mathbf{o}_a^A - \mathbf{o}_b \\ &= \mathbf{o}_b - R_A\mathbf{o}_a^A + r(\cos\gamma, \sin\gamma) + R_A\mathbf{o}_a^A - \mathbf{o}_b \\ &= r(\cos\gamma, \sin\gamma).\end{aligned}$$

When the circles touch on the outside, the outward normal angles of a and b at the point of tangency are $\gamma + 180°$ and γ. Contact occurs when each angle is in the angular extent of its segment: $\gamma + 180° \in [\alpha_a^A + \theta_A, \beta_a^A + \theta_A]$ and $\gamma \in [\alpha_b, \beta_b]$. The contact curve appears as a point at $\theta_A = \alpha_b - \beta_a^A + 180°$ where the end angle of the a interval equals the start angle of the b interval. It grows in angular extent, then shrinks to a point and vanishes at $\theta_A = \beta_b - \alpha_a^A + 180°$ where the start angle of the a interval equals the end angle of the b interval. When one circle touches the other on the inside, the outward normal angles of a and b at the point of tangency are γ and $\gamma + 180°$, but the patch equations are the same.

Figure 3.16a shows a typical case. Segment a has a center $\mathbf{o}_a^A = (1, 0)$, radius $r_a = 0.5$, start angle $\alpha_a^A = -135°$, and end angle $\beta_a^A = -45°$. Segment b has a center $\mathbf{o}_b = (1, 1)$, radius $r_b = 1$, start angle $\alpha_b = 0°$, and end angle $\beta_b = 180°$. The tangency curve is a circle of radius 1.5 centered at $\mathbf{o} = (1 - \cos\theta_A, 1 - \sin\theta_A)$. Contact occurs for $\theta_A \in [-135°, 135°]$. The displayed configurations are $\mathbf{s} = (-1, 0.5, 0°)$, $\mathbf{e} = (1, 0.5, 0°)$, $\mathbf{c} = (-0.5, 1.1, -90°)$, and $\mathbf{d} = (0.5, 1.1, 90°)$. Figure 3.16b–c shows the parametric contact curve and the contact surface.

Fixed-Axis Translation We derive the contact curves of an x translation-y translation pair from the $\theta_A = 0$ cross-section of a general pair. The configuration $(x_A, y_A, 0)$ of the general pair maps to the configuration (x_A, y_B) with $y_B = -y_A$ of the fixed-axis pair. The fixed-axis contact curve is thus obtained by reflecting the $\theta_A = 0$ cross-section around the x_A axis.

3.4 General Boundary Segments

This section studies contacts between general boundary segments. Figure 3.17 shows two examples. The example in figure 3.17b shows that general segments can have simultaneous contacts, whereas simple features cannot. We derive contact equations for general segments in the same way as for simple features. These equations define the contact curves and surfaces of the general segments. Unlike simple features, the equations lack closed-form solutions, so numerical solutions are required. We discuss numerical solution strategies.

The contact conditions are that the two segments share a boundary point and have opposite outward normals there. Let a be a segment on A with equation $f_A(x^A, y^A) = 0$ and vertices $\mathbf{g}_a$ and $\mathbf{h}_a$. Let b be a segment on B with equation $f_B(x^B, y^B) = 0$ and vertices $\mathbf{g}_b$ and $\mathbf{h}_b$. The tangency equations are

$$f_A(p_x^A, p_y^A) = 0$$

$$f_B(q_x^B, q_y^B) = 0$$

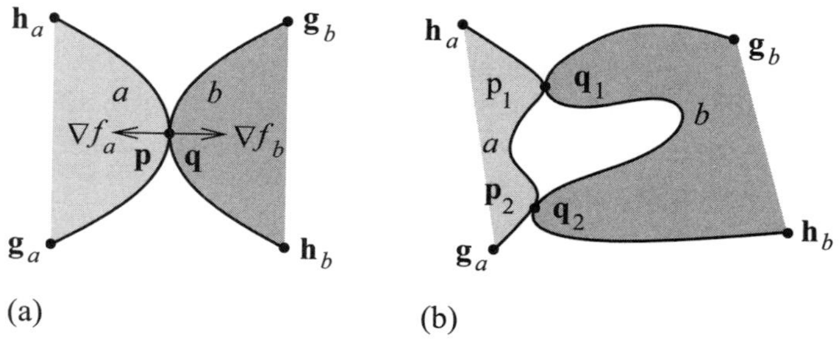

Figure 3.17
General segment contact: (a) quadratics, (b) splines.

$$\mathbf{t}_A + R_A\mathbf{p}^A = \mathbf{t}_B + R_B\mathbf{q}^B$$

$$(R_A\nabla f_A) \times (R_B\nabla f_B) = 0,$$

with $\nabla f = (\partial f/\partial x, \partial f/\partial y)$. The first two equations state that $\mathbf{p}$ and $\mathbf{q}$ are on a and b, the third equation equates these points, and the fourth equation makes the outward normals colinear. Contact occurs when $\mathbf{p}$ and $\mathbf{q}$ lie on a and b:

$$p_x^A \in [g_{ax}^A, h_{ax}^A], \quad p_y^A \in [g_{ay}^A, h_{ay}^A], \quad q_x^B \in [g_{bx}^B, h_{bx}^B], \quad q_y^B \in [g_{by}^B, h_{by}^B].$$

The inequalities assume that a and b are monotonic with respect to x and y. If not, they are split into monotonic subsegments.

As an example, we derive the equations for the two quadratics in figure 3.17a. Segment a has equation $x^A + (y^A)^2 = 0$, vertices $\mathbf{g}_a$ and $\mathbf{h}_a$, and vertex coordinates $\mathbf{g}_a^A = (-1, -1)$ and $\mathbf{h}_a^A = (-1, 1)$. Segment b has equation $x^B - (y^B)^2 = 0$, vertices $\mathbf{g}_b$ and $\mathbf{h}_b$, and vertex coordinates $\mathbf{g}_b^B = (1, 1)$ and $\mathbf{h}_b^B = (1, -1)$. The tangency equations are

$$p_x^A + (p_y^A)^2 = 0$$

$$q_x^B - (q_y^B)^2 = 0$$

$$\mathbf{t}_A + R_A\mathbf{p}^A = \mathbf{t}_B + R_B\mathbf{q}^B$$

$$R_A[1, 2p_y^A]^t \times R_B[1, -2q_y^B]^t = 0.$$

Each segment is split at $(0, 0)$ into two monotonic subsegments. The contact inequalities for the two upper subsegments are

$$p_x^A \in [-1, 0], \quad p_y^A \in [0, 1], \quad q_x^B \in [0, 1], \quad q_y^B \in [0, 1]$$

and likewise for the other three pairs of subsegments.

For a fixed-axis pair, we have five scalar equations in six unknowns: two degrees of freedom and the four components of the two contact points. The solution is a curve in a five-dimensional space whose projection into the configuration space is the contact curve. In our example, suppose a rotates and b translates, so the degrees of freedom are θ_A, x_B, and the constant configuration variables are $x_A, y_A, y_B, \theta_B = 0$. The equations are

$$p_x^A + (p_y^A)^2 = 0$$

$$q_x^B - (q_y^B)^2 = 0$$

$$p_x^A \cos\theta_A - p_y^A \sin\theta = x_B + q_x^B$$

$$p_x^A \sin\theta_A + p_y^A \cos\theta = q_y^B$$

$$(4p_y^A q_y^B - 1)\sin\theta_A - 2(p_y^B + q_y^B)\cos\theta_A = 0$$

and the unknowns are θ_A, x_B, p_x^A, p_y^A, q_x^B, q_y^B. The solution shown in figure 3.17a is where all six unknowns equal zero. For a general planar pair, we have three degrees of freedom, six equations in seven unknowns, and the solution is a surface in a six-dimensional space whose projection is the contact surface.

We need to approximate the contact curve or surface to a specified accuracy. The first step is to solve the tangency equations. If we pick values for all but one degree of freedom, we can solve for the last one by Newtonian iteration. The iteration requires a starting point, which can be hard to derive, and can fail near multiple roots. The approximate contact curve or surface consists of the numerical solution points that satisfy the contact inequalities.

3.5 Spatial Pairs

Spatial contact is more general than planar contact. There are three types of boundary features: faces, edges, and vertices. Each combination of two feature types yields a contact type. It suffices to analyze face-face, face-edge, and edge-edge contacts. The proof that the other cases are covered by this analysis is similar to that for the planar case (section 3.13). The analysis follows the planar pattern: derive tangency equations, solve for the tangency configurations, and use boundary configurations to delimit the contact set.

We derive the equations for face-face, face-edge, and edge-edge contacts. The equations are much harder to solve than in the plane because the feature dimension grows from two to three, the number of degrees of freedom for a part grows from three to six, and the boundaries change from points to curves. For fixed-axis pairs, we have developed closed-form solutions for planar, spherical, and cylindrical faces with linear and circular edges.

Face-Face The tangency equations are the same as for general planar features, except that the points, vectors, and rotations are three-dimensional. The additional condition for face contact is that the contact points lie on the faces. The structure of the contact inequalities reflects that of the face boundaries.

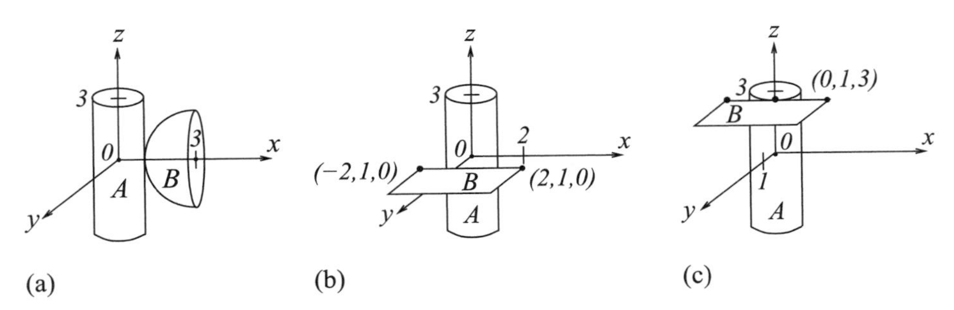

Figure 3.18
Spatial contacts: (a) cylindrical face-spherical face contact, (b) cylindrical face-linear edge contact, (c) circular edge-linear edge contact.

Figure 3.18a shows a contact between a cylindrical face on A with equation $x_A^2 + y_A^2 = 1$ and a spherical face on B with equation $x_B^2 + y_B^2 + z_B^2 = 4$. The configurations of A and B are $(0,0,0,0,0,0)$ and $(3,0,0,0,0,0)$. The tangency equations are

$$(p_x^A)^2 + (p_y^A)^2 = 1$$

$$(q_x^B)^2 + (q_y^B)^2 + (q_z^B)^2 = 4$$

$$\mathbf{t}_A + R_A\mathbf{p}^A = \mathbf{t}_B + R_B\mathbf{q}^B$$

$$R_A[2p_x^A, 2p_y^A, 0]^t \times R_B[2q_x^B, 2q_y^B, 2q_z^B]^t = 0.$$

The cylindrical face is bounded by the half-spaces $z_A \geq -5$ and $z_A \leq 5$, which yield the contact inequalities $-5 \leq p_z^A \leq 5$. The spherical face is bounded by the half-space $x_B \geq 0$, which yields the contact inequality $q_x^B \geq 0$.

Face-Edge Let a be a face on A with equation $f_A(x^A, y^A, z^A) = 0$ and let b be an edge on B with equations $f_B(x^B, y^B, z^B) = 0$ and $g_B(x^B, y^B, z^B) = 0$. Contact occurs when there exist points $\mathbf{p}$ on a and $\mathbf{q}$ on b that coincide with an a normal that is perpendicular to the b tangent. The tangency equations are

$$f_A(p_x^A, p_y^A, p_z^A) = 0$$

$$f_B(q_x^B, q_y^B, q_z^B) = 0$$

$$g_B(q_x^B, q_y^B, q_z^B) = 0$$

$$\mathbf{t}_A + R_A\mathbf{p}^A = \mathbf{t}_B + R_B\mathbf{q}^B$$

$$R_A\nabla f_A \cdot R_B(\nabla f_B \times \nabla g_B) = 0.$$

The structure of the inequalities in face contact is as before. The edge inequalities state that $\mathbf{q}$ lies between the two vertices of the edge.

Figure 3.18b shows a contact between the above cylindrical face and a linear edge on B (the thick line) with equations $y^B = 0$ and $z^B = 0$, vertices $(-2,0,0)$ and $(2,0,0)$, and configuration $(0,1,0,0,0,0)$. The contact equations are

$$(p_x^A)^2 + (p_y^A)^2 = 1$$

$$q_y^B = 0$$

$$q_z^B = 0$$

$$\mathbf{t}_A + R_A\mathbf{p}^A = \mathbf{t}_B + R_B\mathbf{q}^B$$

$$R_A[0,1,0]^t \cdot R_B[1,0,0]^t,$$

since the edge's tangent direction is $(0,1,0) \times (0,0,1) = (1,0,0)$. The edge's inequalities are $-2 \leq q_x^B \leq 2$.

Edge-Edge Let a be an edge on A with equations $f_A(x^A, y^A, z^A) = 0$ and $g_A(x^A, y^A, z^A) = 0$, and let b be an edge on B with equations $f_B(x^B, y^B, z^B) = 0$ and $g_B(x^B, y^B, z^B) = 0$. Contact occurs when there exist points $\mathbf{p}$ on a and $\mathbf{q}$ on b that coincide. The tangency equations are

$$f_A(p_x^A, p_y^A, p_z^A) = 0$$

$$g_A(p_x^A, p_y^A, p_z^A) = 0$$

$$f_B(q_x^B, q_y^B, q_z^B) = 0$$

$$g_B(q_x^B, q_y^B, q_z^B) = 0$$

$$\mathbf{t}_A + R_A\mathbf{p}^A = \mathbf{t}_B + R_B\mathbf{q}^B$$

and the edge inequalities are as before. Figure 3.18c shows a contact between the above linear edge and a circular edge on A with equations $(x^A)^2 + (y^A)^2 = 1$ and $z^A = 0$, vertices $(0,-1,0)$ and $(0,1,0)$, and configuration $(2,0,0,0,0,0)$.

Curved Features Curved features are not amenable to closed-form solutions. Numerical solutions appear feasible for fixed-axis pairs because the number of unknowns is moderate. General pairs have many more degrees of freedom and have not been studied.

3.6 Notes

Contacts between features have been extensively studied in the mechanical engineering and robotics literature. Mechanical engineering research addresses the kinematic analysis of pairs of features in permanent contact. It formulates the general contact constraints, but does not provide a general solution method. Closed-form solutions are available only for gear involutes, which induce linear constraints, and for a few specialized cam profiles. They do not cover imperfect involutes or nonstandard gear and cam profiles. Numerical methods are broadly applicable, but are not robust. They can fail to converge, can converge to spurious solutions, and cannot handle singular contact constraints, such as those induced by a cylindrical shaft in a cylindrical hole. Angeles and Lopez-Cajun [4] and Gonzales-Palacios and Angeles [26] describe numerical solutions for the analysis and synthesis of planar and spatial cams that drive standard followers. Others describe numerical solutions for pairs of convex, planar curves [14, 65]. Litvin [51] describes the feature contacts of gear teeth profiles.

The mechanical engineering literature also addresses infinitesimal kinematics, in which instantaneous displacements and contacts of features are studied [31]. Expressions for instantaneous axes and centers of motion, together with contact invariants, are derived for planar and spatial contacts. The drawback of these methods is that they are local to a linear neighborhood of a known contact configuration.

Robotics studies the contact of parts in configuration space for planning robot motions [47, 52]. The research provides contact equations for polygonal [11, 52] and polyhedral [20] features. The research does not provide practical algorithms for curved features, which are the norm in higher pairs, because they are relatively unimportant for planning motion. Collision detection [50] provides local contact information.

We have studied feature contacts for mechanisms within a configuration space for nearly 20 years [36]. We have studied feature contacts for planar fixed-axis pairs [71], general planar pairs [68], and spatial fixed-axis pairs [42]. Other work includes that by Faltings [23, 24].

Appendix B describes the feature contacts covered by HIPAIR. It covers fixed-axis contacts of simple planar features: line and circle segments.

4 Contact of Parts in Configuration Space

In chapter 3 we began to study contact of parts by characterizing contact of features in configuration space. In this chapter we complete the study of contact of parts. A part contact is a feature contact where the parts' interiors are disjoint. Whereas a feature contact is local to a feature pair, a part contact involves every pair since no features can intersect. We must analyze an entire set of feature contacts to characterize the contacts of parts.

In section 4.1, we model contacts of parts by using configuration space partitions. We specialize partitions to fixed-axis pairs in section 4.2 and present a partition algorithm in section 4.3. We do the same for general planar pairs in sections 4.4 and 4.5. In section 4.6 we generalize from pairs to mechanisms. In section 4.7 we relate our approach to the classic theory of mechanisms.

4.1 Partition of Configuration Space

We characterize a kinematic pair by partitioning its configuration space into free, contact, and blocked spaces. The parts are disjoint in free space, in contact in contact space, and overlapping in blocked space. Figure 4.1 illustrates these concepts. Circle A has a center $\mathbf{o}_a = (0,0)$, radius $r_a = 1$, and translates with $\theta_A = 0°$. Circle B has a center $\mathbf{o}_b = (0,0)$, radius $r_b = 2$, and is fixed. The configuration space is (x_A, y_A). The contact space is the circular contact curve with a center $\mathbf{o}_a + \mathbf{o}_b = (0,0)$ and radius $r_a + r_b = 3$; free space is its exterior and blocked space is its interior.

Free space consists of regions. Figure 4.2 shows a pair with two regions. Circle A is as before and circle B is replaced by a box with a circular cavity. Free space consists of an outer region where A is outside B and a circular region where A is inside the B cavity. Part A can move freely in its initial region but cannot reach the other region. It is placed in the initial region by moving perpendicularly to the motion plane.

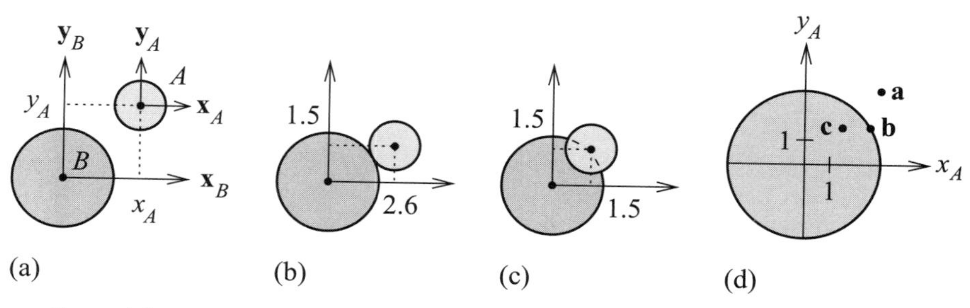

Figure 4.1
Circle-circle pair: (a) free configuration $\mathbf{a} = (3, 3)$, (b) contact configuration $\mathbf{b} = (2.6, 1.5)$, (c) blocked configuration $\mathbf{c} = (1.5, 1.5)$, (d) configuration space partition.

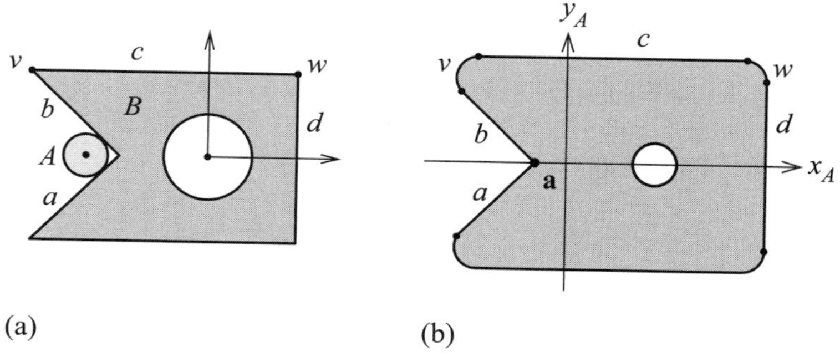

Figure 4.2
Circle-box pair: (a) double contact configuration **a**, (b) configuration space partition.

Contact space contains subspaces where contact occurs between a pair of features. Multiple contacts occur at configurations where the subspaces intersect. In figure 4.2b, the subspaces are curves where the sole A feature touches the b features. The curves for line segments a–d and vertices v–w are labeled. Each curve is part of a feature contact curve, the rest of which is in blocked space. A double contact occurs at configuration **a** where the a and b curves intersect.

Configuration space partitions have a special topological structure. When two parts are free at a configuration, there is a positive minimum distance between pairs of boundary points. If the configuration changes slightly, the boundaries move less than the minimum distance and the parts stay free. Since every free configuration has a neighborhood of free configurations, free space is an open set. Likewise, blocked space is open because blocked parts overlap by a minimum amount and slight changes in configuration

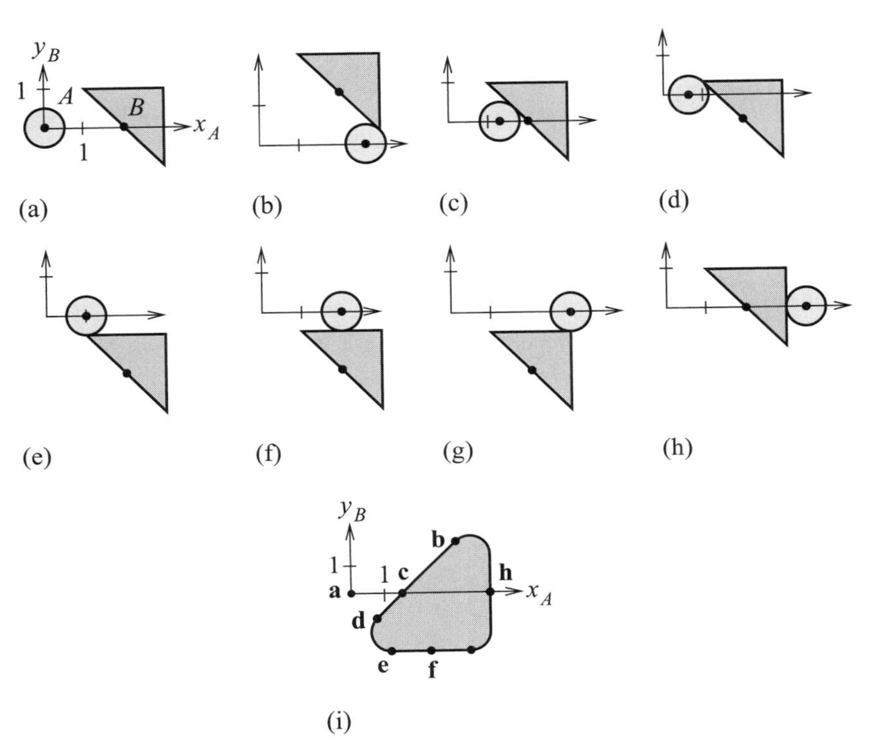

Figure 4.3
Translation pair: (a) configuration $\mathbf{a} = (0, 0)$, (b) $\mathbf{b} = (2.65, 1.35)$, (c) $\mathbf{c} = (1.29, 0)$, (d) $\mathbf{d} = (0.65, -0.65)$, (e) $\mathbf{e} = (1, -1.5)$, (f) $\mathbf{f} = (2, -1.5)$, (g) $\mathbf{g} = (3, -1.5)$, (h) $\mathbf{h} = (3.5, 0)$, (i) configuration space partition.

move the boundaries less than this amount. Contact space is a closed set because it is the complement of the union of these two open sets. At a contact configuration, outward motion along the contact normal frees the parts and inward motion blocks them.

The configuration space partition of a pair provides a global view of its kinematic function that reveals the pairs of features that come in contact, the configurations where the point of contact shifts between feature pairs, and the working modes. We illustrate this view with a simple planar pair (figure 4.3). Circle A translates horizontally with a degree of freedom x_A, and triangle B translates vertically with a degree of freedom y_B. Figure 4.3a–h show configurations **a–h**, while figure 4.3i shows the configuration space partition with these configurations labeled. Configuration **a** lies in the single free region. The contact space geometry encodes the parts' feature contacts. Between configurations **b** and **d**, the circle touches the diagonal edge of the triangle. This corresponds to line segment **bd** in the partition. The contact equation is $y_B = x_A - 1.29$ because the contact curve is parallel to the triangle's edge. It indicates a linear relation between the translations of *A* and *B*.

Between configurations **d** and **e**, the upper left vertex of the triangle follows the circle's boundary, so the corresponding contact curve, **de**, is a circular arc segment. Between configurations **e** and **g** the circle can translate and remain in contact with the horizontal edge of the triangle without pushing the triangle, so the contact curve is a horizontal line segment, **eg**. In this interval, the circle blocks the upward translation of the triangle, which is indicated by the functional relation $y_B = -1.5$. Around configuration **h**, the triangle can translate without pushing the circle and blocks its leftward translation, so the contact curve is a vertical line segment, $x_A = 3.5$. The contact transition configurations **b**, **d**, **e**, and **g** indicate the changes in the kinematic function resulting from changes in contact.

4.2 Fixed-Axis Pairs

A fixed-axis pair has a two-dimensional configuration space. The configuration space is a plane for a translation pair, a cylinder for a rotation-translation pair, and a torus for a rotation-rotation pair. Free space and blocked space are two-dimensional, while the contact space consists of curves.

Figure 4.4 shows a rotating cam that drives a translating follower. Recall the convention that the thick-headed arrow indicates that the cam rotates

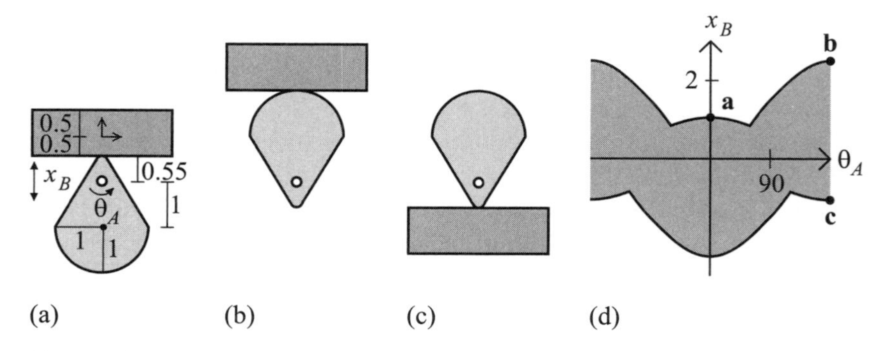

Figure 4.4
Cam pair: (a) configuration $\mathbf{a} = (0°, 1.05)$, (b) configuration $\mathbf{b} = (180°, 2.5)$, (c) configuration $\mathbf{c} = (180°, -1.05)$, (d) configuration space partition.

clockwise and the two-headed arrow indicates that the follower translates up and down. Free space is a single region since the verticals $\theta_A = \pm 180°$ represent one line in the cylinder's configuration space. The contact curves express the vertical distance between the cam's origin and the follower's origin as a function of the cam's angle. For example, $y_A = 0.55 + 0.5 = 1.05$ at configuration **a**, using the displayed dimensions. While the configuration follows a contact curve, the cam rotates and the follower translates with nonlinear coupling between the degrees of freedom.

Figure 4.5 shows a rotating one-tooth ratchet with a rotating pawl. Clockwise rotation of the ratchet is blocked when the pawl tip engages the ratchet tooth (contact transition configuration **a**). Since configuration **a** is a concave point, a clockwise rotation of the ratchet cannot proceed without the pawl rotating clockwise as well. This corresponds to the blocking function of the ratchet-pawl contact. Conversely, contact transition configurations **b** and **c** are convex, so clockwise and counterclockwise rotations of both parts are possible. In this case, the pawl slides over the ratchet tooth. The verticals $\theta_A = \pm 180°$ and the horizontals $\theta_B = \pm 180°$ each represent one line in the torus configuration space.

When a rotating part has rotational symmetry or a translating part has translational symmetry, the configuration space is symmetric in the part's degree of freedom. A part with repeated geometry produces repeated contact curves. For example, a ratchet with four identical teeth yields a configuration space partition with four shifted copies of the one-tooth contact curves (figure 4.6).

Figure 4.7 shows the configuration space partition of the spatial Geneva pair. The free space is a single channel composed of four slanted and four horizontal segments. The channel width determines the mechanical play, which governs the coupling between the cam and the follower. The

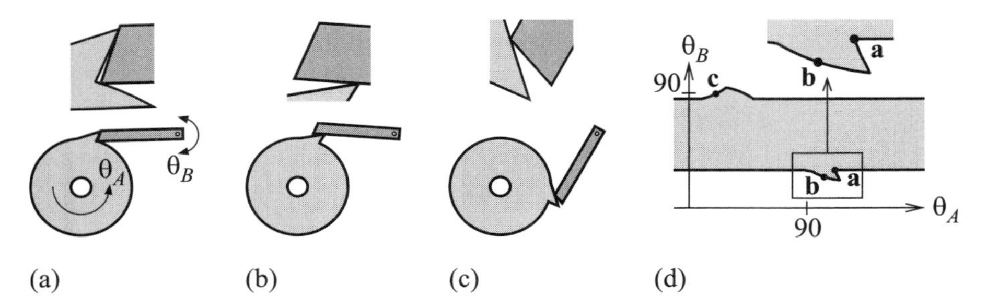

Figure 4.5
Ratchet pair: (a) configuration $\mathbf{a} = (111°, 30°)$, (b) configuration $\mathbf{b} = (103°, 25°)$, (c) configuration $\mathbf{c} = (21°, 89°)$, (d) configuration space partition.

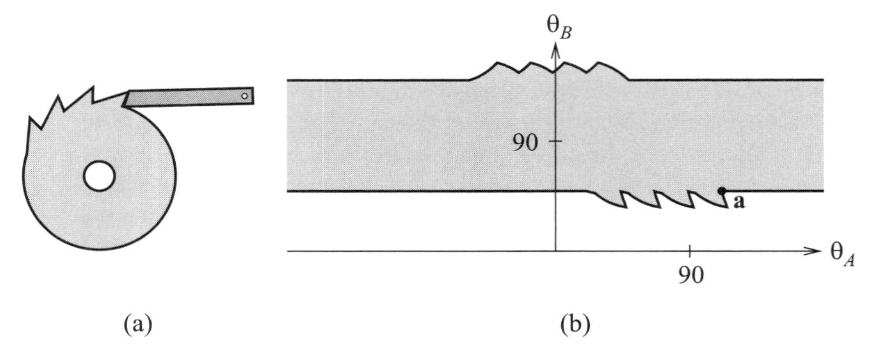

Figure 4.6
Four-tooth ratchet pair at $\mathbf{a} = (111°, 30°)$, (a) configuration, (b) configuration space partition.

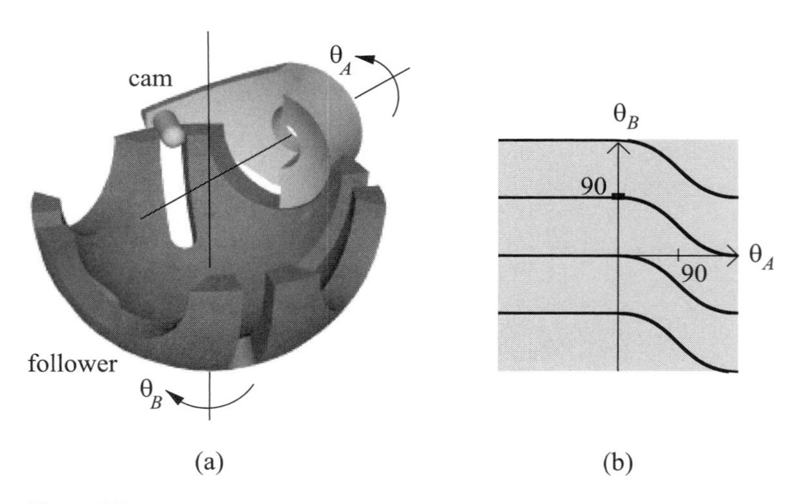

Figure 4.7
(a) A spatial Geneva pair and (b) its configuration space partition.

partition in figure 4.7b shows that the kinematic function is bidirectional and that the cam blocks the follower's rotation in the horizontal segments. The ability to model planar and spatial pairs that have completely different shapes illustrates the generality of configuration space partitions.

4.3 Partition Algorithm for Fixed-Axis Pairs

We construct the partition of a fixed-axis pair from its contact curves. The feature contact curves are augmented with boundary lines that enforce the

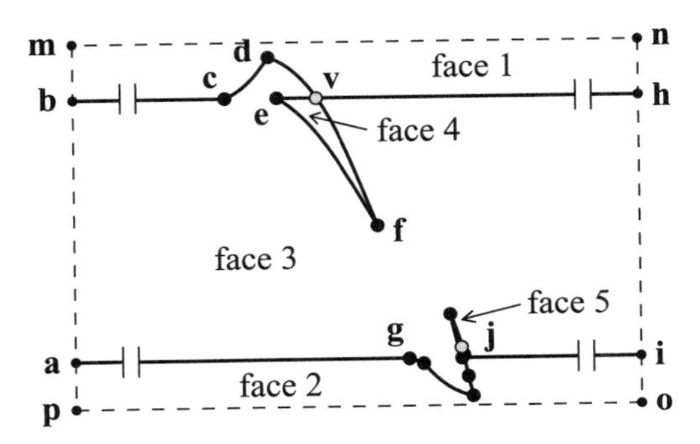

Figure 4.8
Partition structure of a ratchet pair.

Input: part shapes and degrees of freedom.
1. Form contact curves.
2. Split curves into vertices and edges.
3. Form edge loops and faces.
4. Classify faces.
Output: configuration space partition.

Figure 4.9
Partition algorithm for fixed-axis pairs.

topology of the configuration space. These are $\theta_A = \pm 180^\circ$ when θ_A is a degree of freedom and $\theta_B = \pm 180^\circ$ when θ_B is a degree of freedom. The planar region outside the boundary lines is not part of the configuration space. The partition is a planar subdivision of the configuration space.

We split the curves at their intersection points. The resulting curve segments are called edges and their endpoints are called vertices. The complement of the vertices and edges consists of open regions, called faces, that are wholly free or wholly blocked. Each face is bounded by one or more edge loops. We orient the loops so that the face interior lies to the left when they are traversed in order.

Figure 4.8 shows the partition of a one-tooth ratchet pair. Contact curves are solid and boundary lines are dashed. Vertices appear as solid dots for endpoints and gray for intersection points. The edges are **bc**, **cd**, **dv**, **vh**, and so on. Face 1 is bounded by the edge loop **bcdvhnm** and face 4 is bounded by **efv**. Faces 1–2 are free and faces 3–5 are blocked.

Figure 4.9 summarizes the partition algorithm. Step 1 forms the contact curves as described in section 3.2. Steps 2–4 form the vertices, edges, and faces, and classify each face as free or blocked. The next three sections describe these steps.

4.3.1 Vertex and Edge Formation

The first task is to compute the intersection points of n contact curves. One strategy is to test all $n(n-1)/2$ pairs of curves, but this is inefficient when only a few intersect. We expect a constant number of intersections per curve because the curves are short and are distributed evenly in the plane. We employ a standard computational geometry algorithm, called the sweep-line algorithm, whose running time is $O(n \log n)$ in this case. The second task is to form the vertices and edges.

Sweep-Line Algorithm This algorithm sweeps a vertical line along the x axis and tracks the y order of the curves that it crosses. The order is stored in a sweep list. The list is updated when the sweep line reaches a curve's endpoints and intersection points. A curve is inserted at its left endpoint and is removed at its right endpoint. Two curves are swapped in the list at their intersection point. When two curves become adjacent owing to an update, they are tested for intersection.

We illustrate the algorithm on the ratchet pair (figure 4.8). Curve **ag** is inserted in the empty sweep list. Curve **bc** is inserted above **ag** and they are tested for intersection. Curve **bc** is removed from the sweep list, **cd** is inserted above **ag**, and they are tested. Curve **cd** is removed, **df** is inserted above **ag**, and they are tested. Curve **ef** is inserted between **ag** and **df**, and both pairs are tested. Curve **eh** is inserted between **ef** and **df**, both pairs are tested, and intersection point **v** is found. Curves **eh** and **df** swap at **v**, so the sweep list becomes (**ag**, **ef**, **df**, **eh**). Curves **bc** and **eh** are never tested because they are disjoint in x. Curves **eh** and **ag** are never tested because they are separated by other curves at every x.

The algorithm is implemented as follows. The events are stored in a priority queue ordered by x. The queue is initialized with an insert and a remove event for each curve. The first event is dequeued and handled until the queue is empty. Handling updates the sweep list, checks for intersections between newly adjacent curves, and enqueues swap events when intersections occur. Vertical curves, curves that overlap over an interval, or three curves that intersect at a single point are treated as special cases that we do not describe here.

The sweep list is represented as a balanced binary tree. A curve is inserted in the standard way: compare its y with the the root curve y at the current x, recurse on the appropriate subtree, and insert the curve when a leaf is reached. Finding the curves before and after a curve, removing a curve, and swapping two curves are standard. Each operation performs at most $\log n$ comparisons. There are two events per curve and one per intersection,

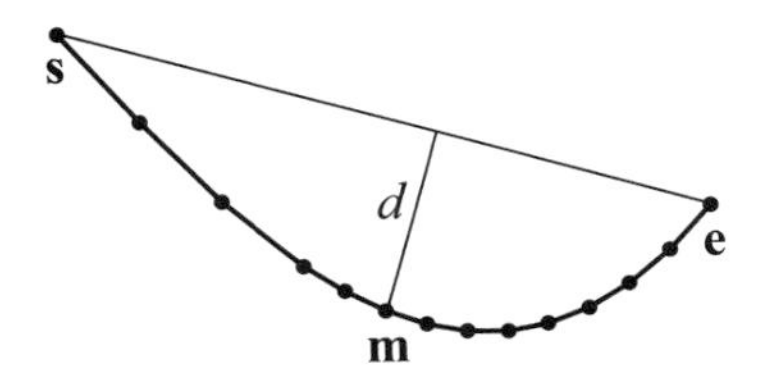

Figure 4.10
Approximation of a contact curve.

which yields order n events, written as $O(n)$, when the number of intersections per curve is constant. Thus, the total running time is $O(n \log n)$.

We compute the intersection points of two curves by intersecting the underlying curves and returning the solutions that lie on both curves. The underlying curves must be intersected numerically because a closed-form solution does not exist, even for line and circle segments. We approximate the contact curves to a specified accuracy using line segments and then intersect the approximate curves.

The approximation algorithm, illustrated in figure 4.10, is as follows. Fit a line through the contact curve's endpoints, **s** and **e**. Sample the curve at $m_x = 0.5(s_x + e_x)$ to obtain the point **m**. If the distance, d, from **m** to the line is less than the specified accuracy, the curve approximation is $(\mathbf{s}, \mathbf{e})$. Otherwise, split the curve at **m**, recursively approximate the two pieces, and concatenate the results. In our example, the initial d is too large, so the curve is split. The sample spacing shrinks from **s** to **e** as the curvature increases.

Edges The edge between two vertices, **p** and **q**, is represented by two oriented edges called twins: **pq** from **p** to **q** and **qp** from **q** to **p**. Edge **pq** is called outgoing at **p** and is called incoming at **q**. Each vertex, **v**, is assigned a list of outgoing edges in clockwise order. The list consists of the twins of the incoming edges in sweep-list order followed by the outgoing edges in reverse sweep-list order. In the ratchet example (figure 4.8), the incoming edges of **v** are **ev** then **dv**, and the outgoing edges are **vh** then **vf**, so the edge list is (**ve**, **vd**, **vh**, **vf**).

4.3.2 Face Formation

The faces are formed by a traversal of the vertex-edge graph. The vertices are traversed in x order; the outgoing edges of each vertex are traversed in clockwise order, and a loop is formed for each untraversed edge. The loop is traversed from its first edge until the first edge is reached again. At each

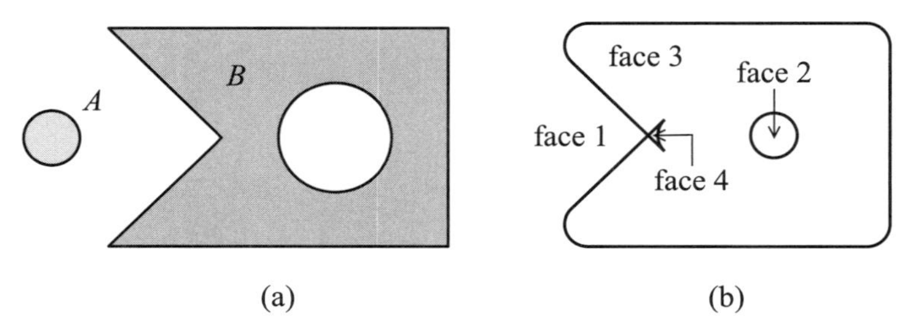

Figure 4.11
(a) Simple pair and (b) its configuration space partition.

step, the next edge is the successor of the twin of the current edge in the clockwise order of its head.

In figure 4.8, vertex **p** is traversed first and yields the loops **pabmnhio** and **poij** $\cdots$ **ga**. Edge **ij** follows **oi** in the second loop because the twin of **oi** is **io** and **ij** follows **io** in clockwise order. Vertex **a** yields one loop, **ag** $\cdots$ **jihvfevdcb**, since its other outgoing edges were traversed in *p* loops. Vertex **b** yields **bcdvhnm**. Vertices **m**, **c**, **d** yield no loops. Vertex **e** yields **efv**. The final loop encloses face 5.

The faces are formed along with the loops. The traversal begins with a single face that is unbounded; hence it has no outer loop. When a vertex is first traversed, its first loop is an inner loop of the enclosing face. Every other loop is the outer loop of a new face. In our example, face 4 has an outer loop **efv**, face 3 has outer loop **ag** $\cdots$ **jihvfevdcb**, and the unbounded face has inner loop **pabmnhio**. Figure 4.11a shows the simple pair from section 4.1 and figure 4.11b shows the four faces in its partition. Face 3 has one outer loop and one inner loop.

4.3.3 Face Classification

It remains to classify each face as free or blocked. We select a configuration in the face, transform the parts to this configuration, and test if they overlap. Two parts overlap if their boundaries intersect or if one is inside the other. The boundaries are tested for intersection with the sweep algorithm. If they are disjoint, we check if one A boundary point is inside B and vice versa. We pick a point, form its upward vertical ray, and count the intersections of the ray with the boundary of the other part. The point is outside the part when the count is even. Figure 4.12 illustrates the test. Classification takes $O(k \log k)$ time for the sweep and $O(k)$ time for the ray test, with k the number of boundary curves.

We greatly reduce the classification time with a fast test during face formation that detects most blocked faces. The contact curve for two features

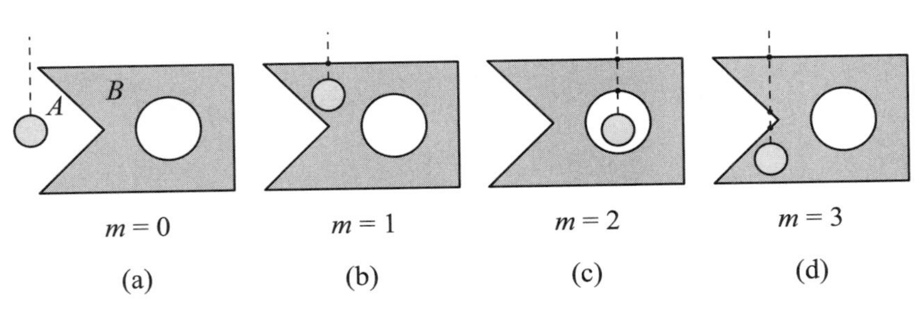

Figure 4.12
Examples of point-in-polygon test: (a) $m = 0$, (b) $m = 1$, (c) $m = 2$, (d) $m = 3$ where m is the intersection count of the vertical dotted line.

is a boundary between configurations where the features are disjoint and where they overlap. We orient the curve so the disjoint configurations are on the left when it is traversed from tail to head. When the curve is split into edges, one twin in each pair, called the positive twin, has the same orientation as the curve, and the other twin, called the negative twin, has the opposite orientation. A face is blocked when its boundary contains a negative edge since the configurations to the left of this edge are blocked and are inside the face.

4.4 General Planar Pairs

General planar pairs have spatial configuration spaces. Free space and blocked space have dimension three. Contact space has dimension two and is a subset of the feature contact surfaces. The $\theta_A = \theta_0$ cross-section of the A/B partition is the planar partition where A translates at orientation θ_0 relative to B. Cross-sections are useful for visualizing and computing spatial partitions.

We illustrate spatial partitions on a fastener pair composed of a pin and a box (figure 4.13). The pin is vertical at $\theta_A = 0°$ (figure 4.13a) and horizontal at $\theta_A = 90°$ (c). The $\theta_A = 0°$ cross-section (figure 4.13b) has two free regions, which indicates that the pin cannot translate out of the box. The $\theta_A = 90°$ cross-section (figure 4.13d) has one free region, which indicates that the pin can translate out of the box. As θ_A decreases from 90°, the clearance between the pin and the opening decreases until the pin touches the top and the bottom of the opening and the cross-section free region splits into two regions.

Figure 4.13e and f show bottom and top views, respectively, of the $\theta_A \in [0°, 90°]$ part of the spatial partition. The $\theta = 0°$ cross-section appears in the bottom view and the $\theta = 90°$ cross-section appears in the top view.

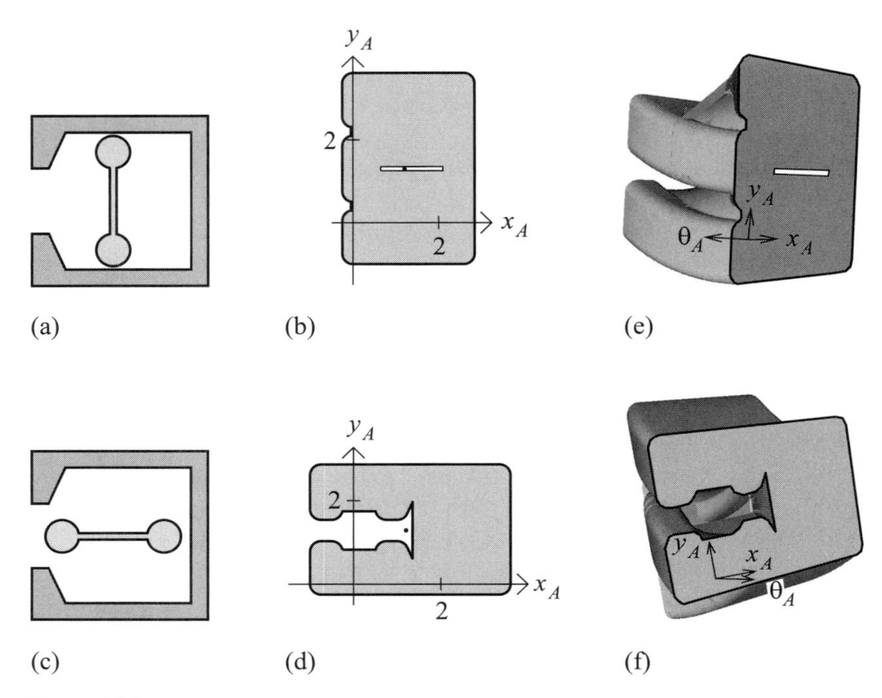

Figure 4.13
Fastener pair: (a–b) $\theta_A = 0°$ cross-section, (c–d) $\theta_A = 90°$ cross-section, (e–f) bottom and top views of spatial configuration space partition.

Although some cross-sections have two free regions, the spatial partition has one free region. The pin can move between any two free configurations by rotating until it is horizontal, translating to the second position, and rotating to the second orientation. The pin can be moved out of the box for $\theta_A \in [46°, 134°]$.

The next example is a constant-breadth cam with a square follower (figure 4.14). When the cam rotates, the follower rotates in step and translates around a circle. In configuration space, the cam moves relative to the follower. Figure 4.14b shows the free region where the cam is inside the follower: a curved channel with a square cross-section. Figure 4.14c shows selected cross-sections projected onto the (x_A, y_A) plane. The center of the square indicates the cam's position when the follower has zero clearance. For the actual clearance, the cam can translate around the nominal position within the square. The clearance is the same regardless of the orientation. A second free region in which the cam is outside the follower lies beyond the blocked region that encloses the channel.

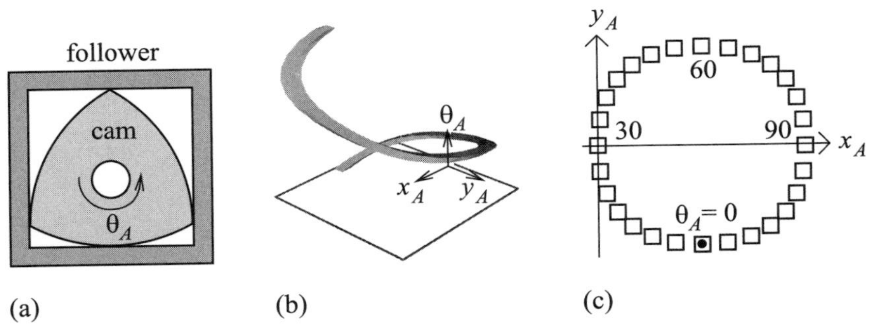

Figure 4.14
Cam pair (a), spatial configuration space partition (b), and projected cross-sections (c).

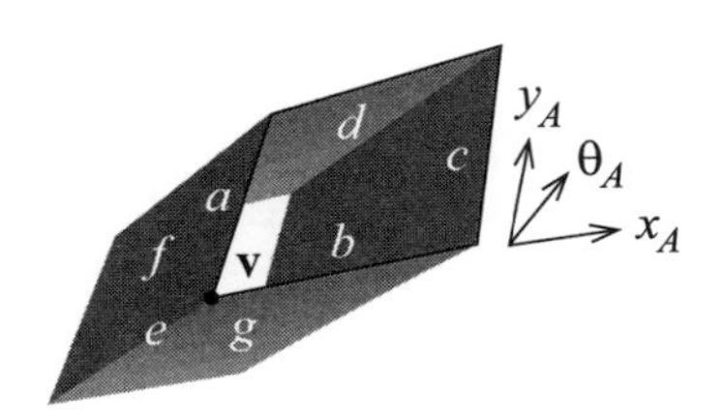

Figure 4.15
Partition structure for a cam pair.

4.5 Partition Algorithm for General Planar Pairs

Spatial partitions are a generalization of planar partitions. The contact surfaces split a configuration space into connected components, called cells, each of which is wholly free or wholly blocked. The surface patches that bound the cells are called faces; their boundary curves are called edges; and the edge endpoints are called vertices. Figure 4.15 shows the $\theta_A \in [0^\circ, 3^\circ]$ part of the cam pair partition. The channel is a cell, f and g are two of its four faces, e is one of its four edges, and there are no vertices.

The spatial partition algorithm (figure 4.16) has the same form as the planar algorithm, form contact surfaces for all pairs of part features, form the cells, faces, edges, and vertices, and classify the cells. The first step is explained in section 4.3. The third step is explained in section 4.3.3. The only change is that the sample configuration is for A relative to B, rather than for A and B relative to the mechanism's frame. It remains to describe the second step.

We compute the spatial partition by computing the $\theta_A = 0^\circ$ planar partition and tracking its evolution as θ_A increases to 360°. As θ_A increases, the

Input: part shapes.
1. Compute contact surfaces.
2. Split surfaces into vertices, edges, and faces.
3. Form cells.
4. Classify cells.
Output: configuration space partition.

Figure 4.16
Partition algorithm for general planar pairs.

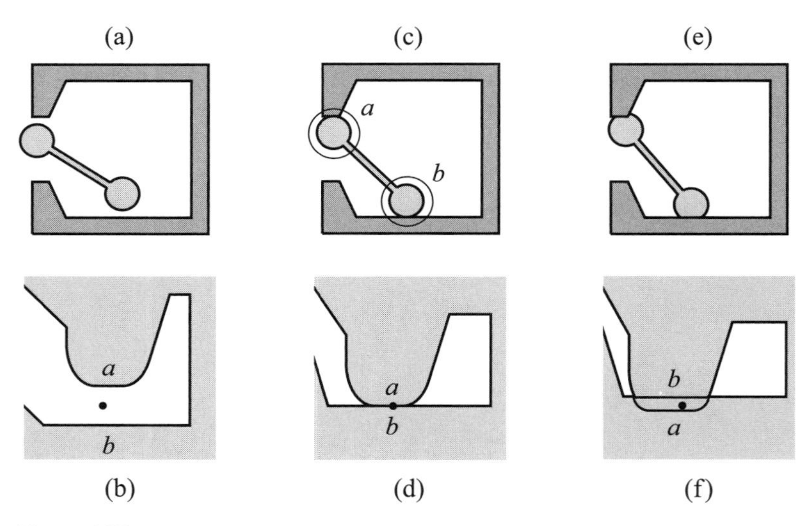

Figure 4.17
Fastener pair configurations and partitions before (a–b), at (c–d), and after (e–f) a contact curve tangency.

faces, edges, and vertices of the planar partition trace the cells, faces, and edges of the spatial partition. In figure 4.15, the planar face bounded by *abcd* traces the channel; planar edges *a* and *b* trace spatial faces *f* and *g*; and planar vertex **v** traces spatial edge *e*. The planar partition is a continuous function of θ_A when its graph is fixed. The contact curves move, which makes the edges move, which makes the vertices move, which makes the faces deform. The graph changes when the edge set or the face set changes.

The edge set changes when θ_A equals the start or end angle of a contact surface, so a curve appears or disappears. Figures 3.14, 3.15, and 3.16 show the three types of contact curves. The edge set also changes when two curves are tangent. Figure 4.17 shows how a tangency changes the free

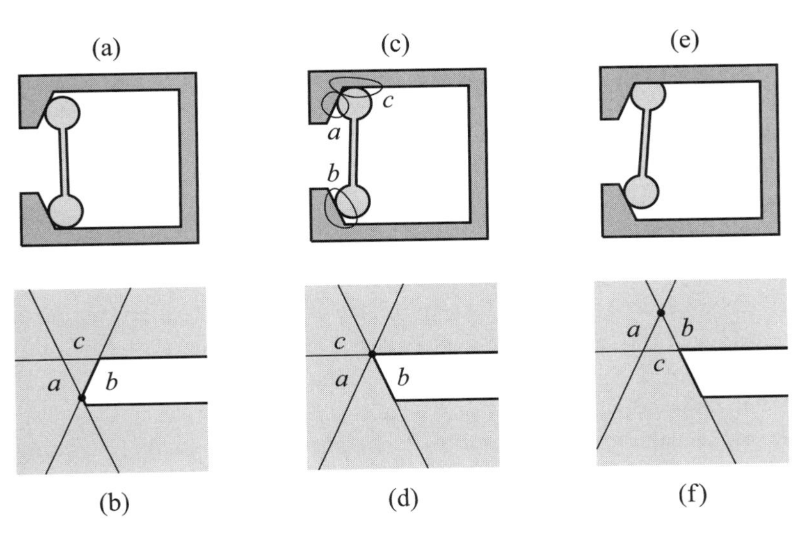

Figure 4.18
Fastener pair configurations and partitions before (a–b), at (c–d), and after (e–f) a face collapses.

space of the fastener pair from one region to two regions. The face set changes when a face collapses to a point, which occurs when three curves intersect. Figure 4.18 shows that a fastener pair face collapses when the upper pin touches the upper slanted box segment (contact *a*), the lower pin touches the lower slanted segment (contact *b*), and the upper pin also touches the upper horizontal segment (contact *c*). Contacts *a–c* are labeled in the triple contact configuration (figure 4.18c) and in the three partitions (figure 4.18b,d,f).

The contact curve's start and end angles are described in section 3.3. Two curves are tangent when the underlying curves are tangent and the point of tangency lies on both curves. Tangency occurs when the parametric expressions from section 3.3 satisfy the contact equations from section 3.1.1. There is one case for each pair of contact types. For example, consider the contact line of moving circle a and fixed line b, and the contact circle of moving circle e and fixed circle f. A line with normal $\mathbf{n}$ and distance from the origin k is tangent to a circle with center $\mathbf{o}$ and radius r when $\mathbf{n} \cdot \mathbf{o} = r + k$. Equations 3.7 and 3.8 yield $\mathbf{n} = \mathbf{n}_b$, $\mathbf{o} = \mathbf{o}_f - R_A \mathbf{o}_e^A$, $r = |r_e \pm r_f|$, and $k = r_a + d_b - \mathbf{n}_b R_A \mathbf{o}_a^A$. The resulting equation has a closed-form solution. The same is true of the other cases.

Three curves intersect when the underlying curves intersect and the intersection point lies on all three curves. The intersection configurations are the solutions of the three parametric contact curves. There is one case for each triple of contact types. Numerical solutions are required.

The planar partition is updated at angles where its graph changes. When a curve appears, it is added to the partition, which entails splitting the edges that it intersects and updating the vertices and faces. When a curve disappears, it is removed from the partition, which entails merging edges that it formerly split and faces that it formerly separated. When disjoint curves become tangent and then intersect, the edges that contain the point of tangency are split and the face that they bound splits into two faces (figure 4.17b,d,f). When the opposite occurs, the edges that formerly intersected are merged and two faces merge. When a face collapse occurs, the old face is replaced with a new one.

The spatial partition is constructed from the evolving planar partition. Planar faces, edges, and vertices generate spatial cells, faces, and edges. A spatial vertex is formed when two spatial edges from the same surface meet at its start or end angle, and when two edges from different surfaces meet at a face collapse. Although the construction is straightforward, there are many details and special cases.

4.6 Mechanisms

We have seen that the partition of the configuration space of a kinematic pair models its parts' contacts. We model the contacts among the parts of a mechanism in the same way. The pairwise contacts are modeled by constructing a configuration space partition for each kinematic pair in the mechanism. Contacts among three or more parts are modeled by composing the pair partitions to form a partition of the mechanism's configuration space. The configuration space partition of the mechanism consists of free regions where all the pairs are free, blocked regions where some pairs are blocked, and contact regions where some pairs are in contact and the rest are free. The contact region geometry encodes the coupling among the degrees of freedom of the touching parts.

We illustrate configuration space partitions for a mechanism by using a pawl indexing mechanism composed of a driver D, indexer I, and pawl P (figure 4.19). As the driver rotates clockwise, its finger engages an indexer pin, rotates the indexer counterclockwise by one-tenth of a turn, and disengages. When the driver is disengaged (figure 4.19a), its rim aligns with the left pawl arm, which prevents clockwise rotation, and the right pawl arm prevents indexer rotation by engaging two pins. When the driver engages the indexer, its notch aligns with the left pawl arm, which allows the bot-

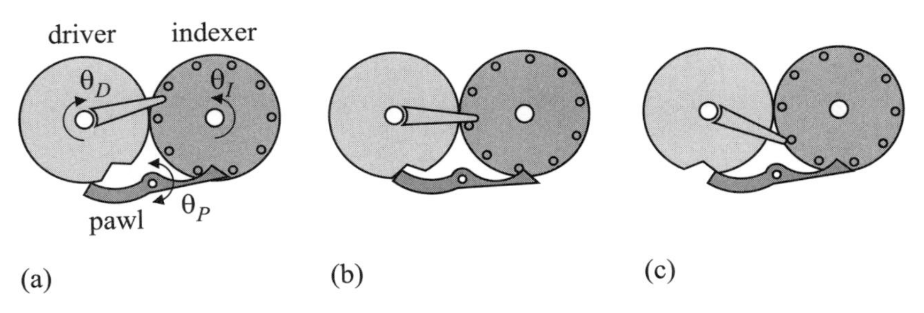

Figure 4.19
Indexing mechanism: (a) disengaged, (b) engaged, (c) disengaging.

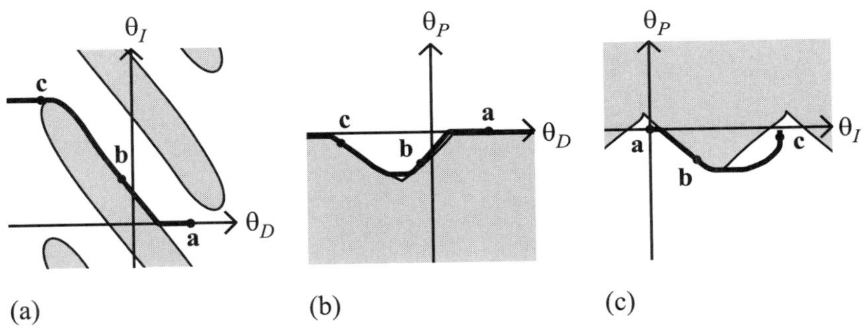

Figure 4.20
Indexing mechanism pair configuration space partitions: (a) driver-indexer (detail), (b) driver-pawl, (c) indexer-pawl.

tom indexer pin to rotate the right pawl arm out of the way (figure 4.19b). When it disengages, the left pawl arm returns to its rim and the right arm re-locks the indexer (figure 4.19c).

Figure 4.20 shows details of the configuration space partitions of the indexing mechanism's pairs. The driver-indexer free space consists of ten diagonal channels where the finger engages the ten pins. The negative slope indicates that the parts rotate in opposite directions. The rest of the free space is the regions to the left and right of the channels where the driver is disengaged. The driver-pawl free space is bounded from below by the rim-left arm contact curve, which is horizontal, and by the notch-left arm contact curves, which form a v-shape (figure 4.20b). The horizontal curve blocks downward vertical motion, clockwise pawl rotation, whereas the slanted curves permit it. The indexer-pawl free space (figure 4.20c) is bounded from above by the pin-right arm contact curves, which form ten inverted v-shapes. When θ_I increases or decreases, θ_P decreases along one

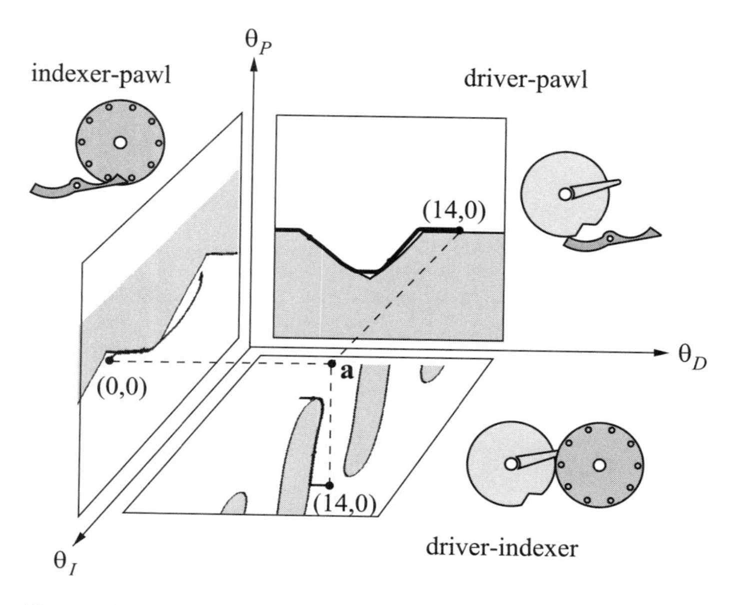

Figure 4.21
Illustration of the configuration space partition of the indexer mechanism as a composition of its pair configuration space partitions.

side of the v-shape. In other words, indexer rotation in either direction causes clockwise pawl rotation.

The kinematic function of a mechanism involves simultaneous contact of parts. For example, the driver is in contact with the indexer and the pawl in configuration **a**, whereas the indexer and the pawl are not in contact (figure 4.19a and configuration **a** in figure 4.20). As the driver rotates clockwise, the indexer and the pawl also come into contact (figure 4.19b and configuration **b** in figure 4.20). Partition of the mechanism's configuration space encodes this contact information.

Figure 4.21 illustrates partition of the indexer's configuration space. We focus on a small region because the full partition is difficult to visualize. Each configuration in the partition of the pair's configuration space is extended to a line parallel to the third configuration space parameter in partition of the mechanism's configuration space. The intersection of the three lines corresponds to the mechanism's configuration space. For example, configuration $\mathbf{a} = (14^\circ, 0^\circ)$ in the driver-pawl pair configuration space is replaced by the line $(14^\circ, 0^\circ, \theta_I)$; configuration $(14^\circ, 0^\circ)$ in the driver-indexer pair configuration space is replaced by the line $(14^\circ, \theta_P, 0^\circ)$; and configuration $(0°, 0°)$ in the indexer-pawl pair configuration space is replaced by the line $(\theta_D, 0°, 0°)$. They intersect at configuration $\mathbf{a} = (14°, 0°, 0°)$ of the mechanism.

The computation of configuration space partitions for mechanisms is an active research topic in robotics and in computational geometry. One approach employs set operations. Free space is the intersection of the free spaces of all the pairs because a configuration is free when every pair is free. Blocked space is the union of pair blocked spaces because a configuration is blocked when any pair is blocked. Contact space is the complement of the union of free space and blocked space. In set operations, the pair configuration $\mathbf{a}$ is replaced by the set of mechanism configurations $(\mathbf{a}, \mathbf{b})$ where $\mathbf{b}$ ranges over the degrees of freedom of the other parts.

The computational complexity of computing a configuration space partition has been proved to grow exponentially with the number of a mechanism degrees of freedom. Even a single spatial pair, which has six degrees of freedom, appears unmanageable. Although computing the entire partition is impractical, partial computation is a valuable design tool. The simplest case is to construct a mechanism contact region from a set of pair contact regions. For kinematic analysis, the partitions of the pair configuration space provide insights into the individual workings of the pairs and their role in the mechanism's kinematic function.

4.7 Partitions of Configuration Space and the Theory of Mechanisms

The configuration space model of contact of parts is an extension of the classic models in the theory of mechanisms. The theory focuses on linkage, cam, and gear mechanisms. In each case, the mechanism is modeled under restrictive assumptions about the parts' contacts. The classic models are equivalent to the contact spaces in the configuration space models of the mechanisms.

Cams A cam is assumed to be in permanent contact with its follower, with the follower's position a function of the cam's orientation. In our cam example (figure 4.4), x_B is the function of θ_A given by the lower boundary of the upper free space region. Intermittent contact is not modeled. We treat a cam and its follower as a higher pair, so permanent contact and intermittent contact are modeled.

Gears Gears are assumed to be in permanent contact. Normally the contact is via involute teeth and each gear has one degree of freedom, so the

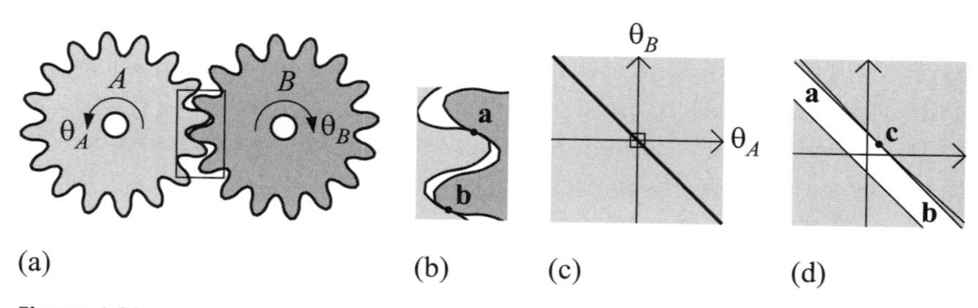

Figure 4.22
Gear pair with detail (a–b) and partition with detail (c–d).

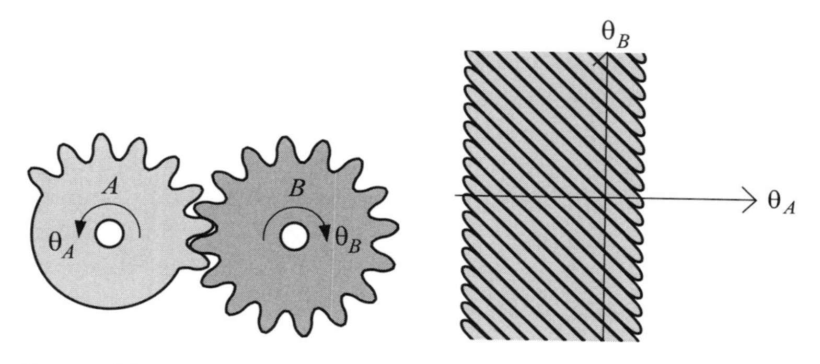

Figure 4.23
Half-gear pair and its partition.

mechanism has one degree of freedom. Imperfect teeth, clearance, undercutting, hub play, and other changes in contact are not modeled. We treat gears as higher-pair mechanisms. Gears with hub play have three degrees of freedom, as do floating gears. Other gears are modeled as fixed-axis parts.

Figure 4.22 shows a fixed-axis gear pair in which the involutes are approximated with circle segments. As gear A rotates counterclockwise, the contact point shifts from feature pair a to feature pair b at configuration **c** (figure 4.22b,d). The **a** and **b** contact curves are nonlinear and intersect transversely at **c**, whereas true involute contact curves are parallel line segments that meet at **c**. The free space represents clearance between the gears. When A is stationary, B can rotate by the vertical extent of the free space. True involutes have zero clearance, so the free space is empty.

The standard gear model does not cover sector gears, such as the half-gear pair in figure 4.23. Meshing and unmeshing involve contact changes between noninvolute tooth features. Partitions model uncoupled motion,

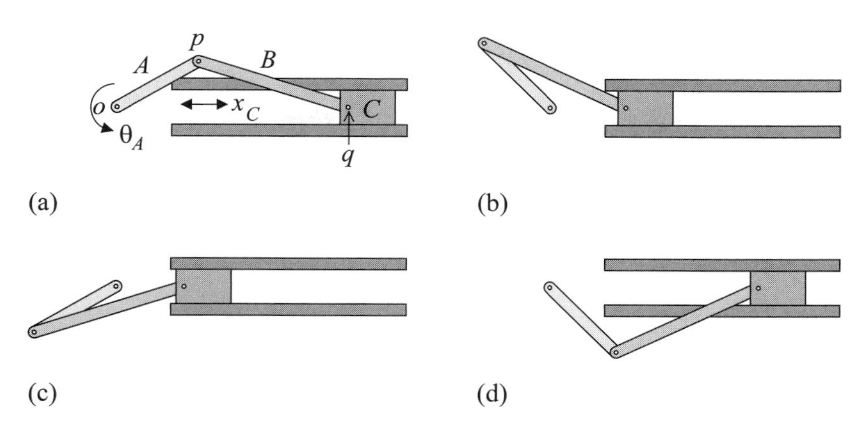

Figure 4.24
Crank-and-slider configurations: (a) $\mathbf{a} = (30°, -17°, 2.9)$, (b) $\mathbf{b} = (135°, -25°, 1.0)$, (c) $\mathbf{c} = (-150°, 17°, 0.9)$, (d) $\mathbf{d} = (-45°, 25°, 2.7)$.

coupled motion, and contact changes. In our example, the free space consists of 16 diagonal channels where the gears mesh, surrounded by a region where they disengage. The 16 channels represent the 16 teeth of B that can mesh with the lead tooth of A.

Linkages A linkage is assumed to consist of links connected by joints. The configuration variables satisfy equality equations whose parameters are the link lengths and the joint configurations. Most linkages have one degree of freedom, so all the variables are determined once a single variable is known. Joint limits, joint play, and link interference are not modeled. We treat a linkage as an assembly of lower and higher pairs. Lower pairs model the joints for which the classic assumptions are acceptable. Higher pairs model the other part contacts.

We illustrate linkage partitions on a three-link crank-and-slider mechanism (figure 4.24). Link A is attached to the frame by a revolute joint at $o = (0, 0)$ and link C is attached to the frame's x axis by a prismatic joint. Link B is attached to A and C by revolute joints at p and q. The link's coordinate frames are placed at o, p, and q with the x axes parallel to the links. The degrees of freedom are θ_A, x_B, y_B, θ_B, x_C with $x_A, y_A, y_C, \theta_C = 0$. The contact equations are

$$l_A \cos \theta_A = x_B$$

$$l_A \sin \theta_A = y_B$$

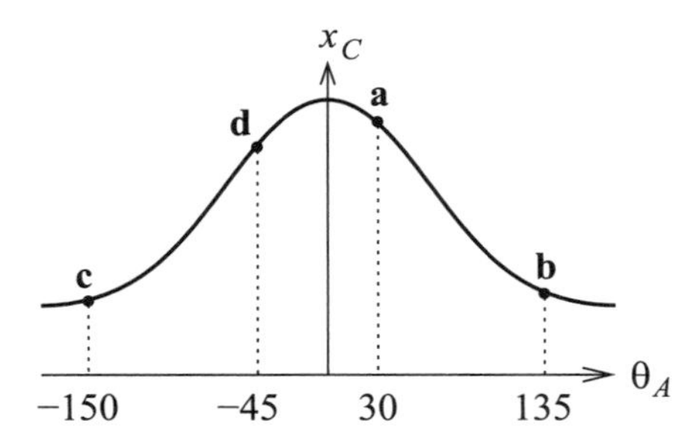

Figure 4.25
Kinematic function of a crank and slider.

$$x_B + l_B \cos\theta_B = x_C$$

$$y_B + l_B \sin\theta_B = 0,$$

with l_A and l_B the link's lengths. These four equations define a curve in the five degrees of freedom. We can eliminate x_B and y_B from the last three equations using the first two equations. The reduced system defines a curve in $(\theta_A, \theta_B, x_C)$ whose projection onto (θ_A, x_C) is the linkage input-output function (figure 4.25).

4.8 Notes

The contact of parts is closely related to the contact of features and has been extensively studied in the mechanical engineering and robotics literature. Mechanical engineering research addresses part contacts for cam, gear, and linkage mechanisms. In cam and gear mechanisms, the contact sequence between parts is predetermined [4, 26, 51], and feature contacts are determined as described in chapter 3. Linkage mechanisms are described as rigid links connected by standard joints with no play.

Robotics studies the contact of parts within the configuration space method for planning robot motion [16, 47, 52]. The research provides algorithms for partitioning configuration space for planar polygons [11, 52] and partial algorithms for polyhedral parts [20]. The research does not address curved parts. Bajaj and Kim [6, 7] describe algorithms for computing a configuration space partition for parts with algebraic curves and surface contours.

For a detailed description of the sweep-line algorithm and related computational geometry tasks, see Berg et al. [18].

Reuleaux's seminal work [67] analyzes mechanisms as collections of lower and higher pairs whose kinematics is derived by composing a pair's kinematics. The configuration space partition of mechanisms introduced in this chapter formalizes this compositional approach.

We have studied the contact of parts in mechanisms within configuration space for nearly 20 years [36]. We have studied part contacts for planar fixed-axis pairs [71], general planar pairs [68], and spatial fixed-axis pairs [42]. Other work includes that by Falting [23, 24].

Appendix B describes the configuration space partitions constructed by HIPAIR. The program handles fixed-axis planar mechanisms.

5 Analysis

The next four chapters explore kinematic design using configuration spaces. This chapter addresses analysis. The purpose of analysis is to determine how a mechanism functions. Kinematic analysis studies the functional consequences of the requirement that the configuration of every mechanism must lie in free space or in contact space. Partitions of the configuration space provide a geometric view of these spaces that facilitates the analysis. The feature contact spaces reveal the structure of the individual contacts, while their intersection spaces reveal the coupling among contacts. The benefit is greatest when seemingly unrelated features interact.

In section 5.1 we discuss kinematic analysis through an examination of configuration space partition. The strategy is to relate the geometric properties of partitions to functional properties and then to compare mechanisms by comparing their geometric properties. We illustrate this strategy with several examples. The remainder of the chapter is devoted to simulation. Simulation predicts the function of a mechanism by solving differential equations that express Newton's laws of motion. In kinematic simulation, the initial values and velocities of the mechanism's degrees of freedom are given, whereas in dynamical simulation the initial state and the external forces are given. Simulation is less general than examination in that many inputs may be required to capture every aspect of the mechanism's function. We present a kinematic simulation algorithm in section 5.2 and extend it to dynamical simulation in section 5.3. The use of configuration space partitions enables us to predict contact changes, whereas prior simulators are limited to mechanisms with permanent contacts or with known contact changes.

5.1 Kinematic Analysis by Examination of Partition

We saw in chapter 4 that an examination of a mechanism's configuration space partitions provides qualitative and quantitative information about its

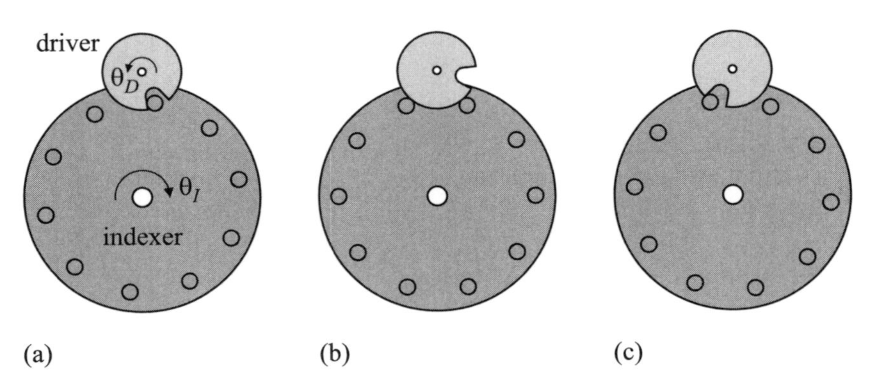

Figure 5.1
Disk indexing pair: (a) drive configuration **a**, (b) dwell configuration **b**, (c) drive configuration **c** in next period.

kinematic function. Kinematic function is typically specified with mixed structural, behavioral, functional, and historical descriptions, such as switching, balancing, engaging, locking, and braking. While a systematic configuration space partition analysis is desirable, it would require a formal specification of kinematic function and a classification of mechanisms into a functional taxonomy, neither of which is available in the mechanical engineering literature. Instead we demonstrate the efficacy of an examination of partition with numerous examples throughout the book, including a catalog of mechanisms (appendix A).

In this section we examine two fixed-axis indexing pairs with similar kinematic functions. Although the part geometries are different, the configuration space partitions are similar. Changing the nominal values of the design parameters leads to changes in kinematic function, including failure modes, that are apparent in the configuration space partitions. We then examine an intermittent gear mechanism with two general planar pairs.

5.1.1 Disk Indexing Pair

The first indexing pair is shown in figure 5.1. The driver is a disk with a slot cutout. The indexer has ten circular pins evenly spaced on a disk and is spring-loaded clockwise. Continual counterclockwise rotation of the driver causes intermittent clockwise rotation of the indexer. In drive periods, the driver slot engages an indexer pin and advances the indexer one-tenth of a turn. In dwell periods, the outer arc of the driver engages two adjacent indexer pins and prevents rotation. When a dwell period ends, the indexer spring rotates the next indexer pin into the driver slot and the next drive period begins.

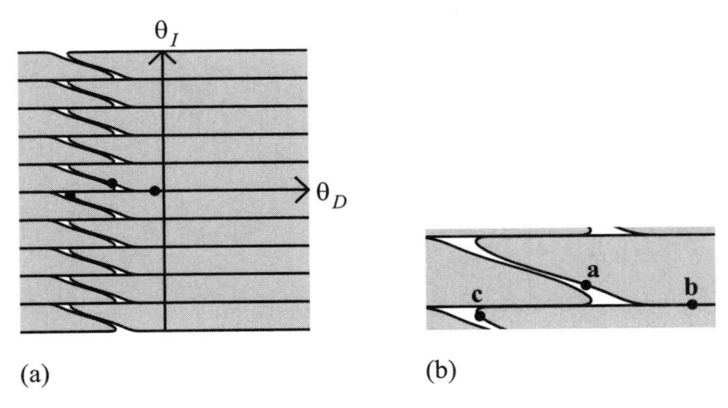

Figure 5.2
(a) Configuration space partition and (b) detail.

Figure 5.2a shows the configuration space partition with the three configurations from figure 5.1. There is one free region with ten slanted segments that represent drive periods and ten horizontal segments that represent dwell periods. The segment width corresponds to the play between the indexer and the driver. The play is largest at transitions between drive and dwell periods. The partition reveals a failure mode. If the spring is too weak, the driver slot will pass the indexer pin before it can engage, so the indexer will not advance (the path in figure 5.2b from configurations **b** to **c** instead of from **b** to **a**).

We present two examples of how changes in the partition reflect changes in kinematic function. Figure 5.3 shows the effect of decreasing the distance between the centers of rotation by 1% from 33 mm to 32.7 mm. The free space splits into ten regions where the driver blocks on the ten indexer pins. The driver can rotate one pin, but cannot disengage from it and rotate to the next pin (figure 5.3c). Figure 5.4 shows the effect of decreasing the width of the driver slot mouth by 15% from 6 mm to 5.1 mm. The free space again contains ten regions, which show a second type of blocking (figure 5.4b). The driver can engage and disengage the indexer pins, but cannot rotate the indexer by one-tenth of a turn (figure 5.4c).

5.1.2 Lever Indexing Pair

The second indexing pair is shown in figure 5.5. The driver consists of a driving lever and a locking arc mounted on a disk (not shown). The indexer has eight pins evenly spaced on a disk. Continual counterclockwise rotation of the driver causes intermittent clockwise rotation of the indexer. In

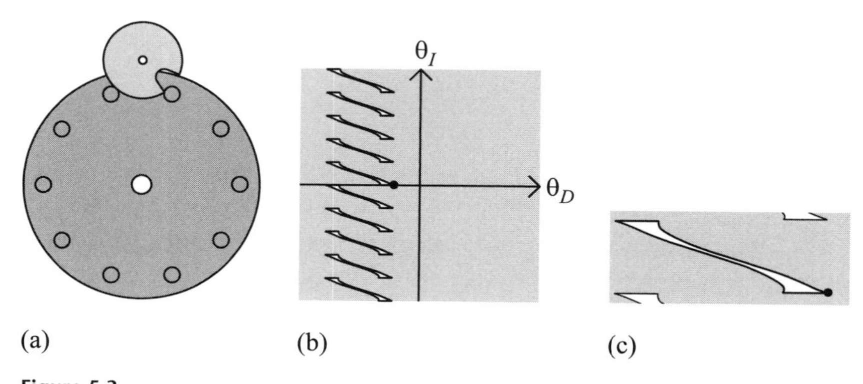

Figure 5.3
(a) Modified disk indexing pair with centers of rotation too close, (b) partition, (c) detail.

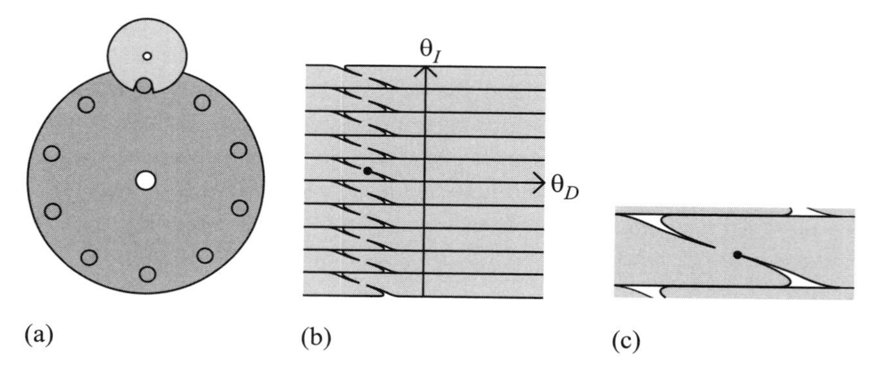

Figure 5.4
(a) Modified disk indexing pair with slot too narrow, (b) partition, (c) detail.

drive periods, the driver lever engages an indexer pin and advances the indexer one-eighth of a turn. In dwell periods, the locking arc engages two adjacent indexer pins and prevents rotation.

Figure 5.6 shows the configuration space partition with the six configurations from figure 5.5. The slanted and horizontal segments in figure 5.6a represent the drive and dwell periods. The intended kinematic function occurs along the path **abcde** in figure 5.6b. The partition reveals two failure modes. The indexer can rotate clockwise from **b** to **f** and block. It can also rotate counterclockwise from **d** to **a** and repeat the previous cycle.

We can remove both failures by modifying the driver (figure 5.7). The first failure occurs when the leading side of the locking arc touches an

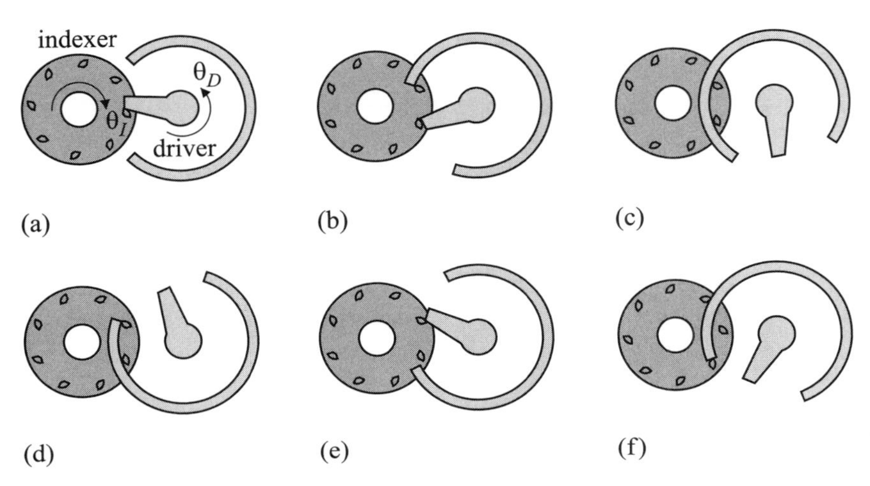

Figure 5.5
Lever indexing pair: (a) drive configuration **a**, (b) transition to dwell at configuration **b**, (c) dwell configuration **c**, (d) end of dwell at configuration **d**, (e) start of next drive period at configuration **e**, (f) blocking configuration **f**.

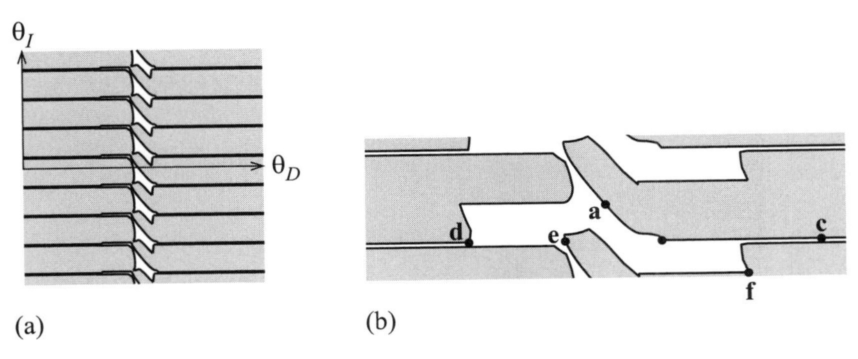

Figure 5.6
(a) Configuration space partition, and (b) detail.

indexer pin and forms a contact curve with a negative slope. As the driver angle increases, the indexer angle decreases until the configuration reaches a blocking configuration (like the ratchet pair in figure 4.5). We assign the contact curve a positive slope, which eliminates the blocking configuration, by beveling the locking arc side. The second failure occurs because the locking arc needs to engage two indexer pins for dwell. It must disengage one pin before the dwell ends, so that it can disengage the second

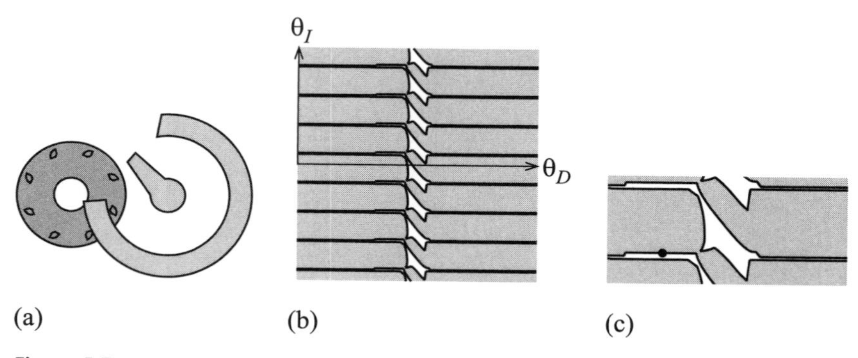

Figure 5.7
(a) Revised lever indexing pair, (b) configuration space partition, and (c) detail.

pin at the start of the next drive period. We prevent indexer rotation by increasing the outer radius of the locking arc so it locks another pin.

The configuration space partition of the modified pair resembles that of the spatial Geneva pair (figure 4.7). Both partitions exhibit slanted driving segments linked by horizontal dwell segments. The number of segments equals the drive ratio of the pair. The Geneva partition has smooth segment boundaries, which indicates smooth motion of parts, whereas the second partition has irregular boundaries. The partition of the first indexer pair resembles these partitions since all three pairs have the same kinematic function, but it is also spanned by horizontal segments, owing to its failure mode (figure 5.1).

5.1.3 Intermittent Gear Mechanism

General planar pairs have more complex kinematic functions than fixed-axes pairs because of their extra degree of freedom. Direct examination of the three-dimensional partition of the configuration space is practical in relatively simple pairs, such as the constant-breadth cam pair in section 4.4. Another strategy is to examine partition cross-sections, which represent translation of a part at a fixed orientation.

We illustrate cross-section examination using the pawl indexer gear mechanism in figure 5.8. The mechanism consists of a constant-breadth cam, a follower with two pawls, and a gear with inner teeth. The cam and the gear are mounted on a fixed frame and rotate around their centers with degrees of freedom θ_C and θ_G; the follower is free with degrees of freedom (x_F, y_F, θ_F). Rotating the cam causes the follower to rotate in step while reciprocating along its length. The right follower pawl engages a gear tooth at cam angle $\theta_C = 0°$ (figure 5.8a); the follower rotates the gear 45°; the

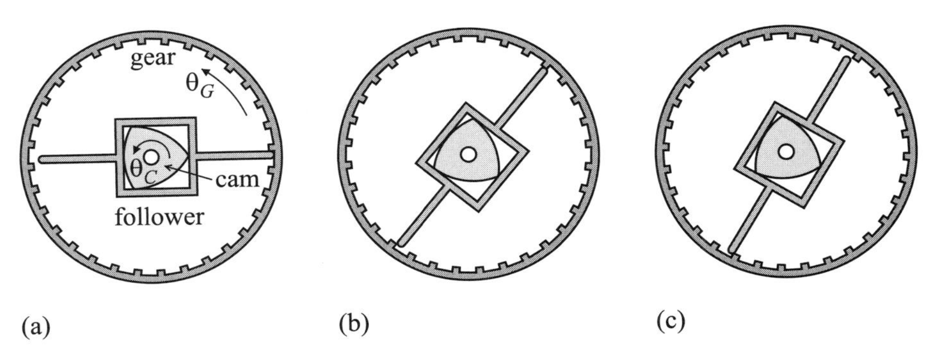

Figure 5.8
Intermittent gear mechanism: (a) right follower pawl engaged, (b) follower disengaged, (c) left pawl engaged.

right pawl disengages at $\theta_C = 82°$ (figure 5.8b); the follower rotates independently while the gear dwells; then the left pawl engages the gear at $\theta_C = 90°$ and the cycle repeats (figure 5.8c).

Figure 5.9 shows the cross-sections of the gear-follower partition that correspond to the snapshots. The follower's configuration is $(0, 0, 0)$ and the gear degrees of freedom are (x_G, y_G). In figure 5.9a, the free space consists of three regions. In the middle region, which contains the snapshot configuration, the right follower pawl engages a gear tooth. The gear motion is about 10 mm, whereas its outer radius is 175 mm. In the region above or below, the pawl engages the tooth below or above. In figure 5.9b, the three regions have merged into a single region, which shows that the follower can now cross over the two gear teeth. In figure 5.9c, the free space once again consists of three regions and the middle region contains the snapshot configuration in which the left pawl engages a gear tooth.

5.2 Kinematic Simulation

Simulation is a method of computing the state of a time-varying system. The state at time t is $\mathbf{x}(t)$, it initial value is $\mathbf{x}(t_0)$, and its time derivative is $\dot{\mathbf{x}}(t)$. The system satisfies a system of ordinary differential equations, $\dot{\mathbf{x}} = \mathbf{f}(\mathbf{x}, t)$. For example, $\dot{x} = -2tx^2$ with $x(0) = 1$ yields $x(t) = 1/(1 + t^2)$. Although few systems have closed-form solutions, there are many efficient numerical solvers that generate sequences of approximate solution points $\mathbf{x}(t_0), \mathbf{x}(t_1), \ldots, \mathbf{x}(t_n)$. Each point is derived by a formula that contains earlier points and $\mathbf{f}$ values. The simplest formula, Euler's method, is $\mathbf{x}(t + h) = \mathbf{x}(t) + h\mathbf{f}(\mathbf{x}(t), t)$ with h a small constant. Better formulas employ more points and $\mathbf{f}$ values.

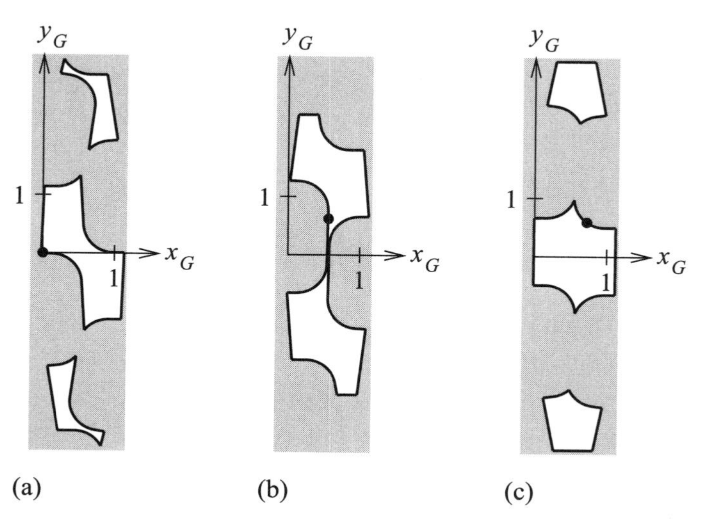

Figure 5.9
Gear-follower partition cross-sections: (a) $\theta_G = 0°$, (b) $\theta_G = 5°$, (c) $\theta_G = 9°$.

The state of a mechanism is its configuration. Kinematic simulation assumes that the derivative is a function of a driving motion: a velocity for one or more parts over a time interval. The parts' velocities are determined by propagating the driving motion through the parts' contacts. Parts in contact are assigned velocities that make the relative velocity at the contact point tangent to the parts' boundaries. Equivalently, the velocity vector is tangent to the contact space.

Figure 5.10 shows a kinematic simulation of the figure 4.3 pair. Circle A translates horizontally with a degree of freedom x_A and triangle B translates vertically with a degree of freedom y_B. The state is $\mathbf{x} = (x_A, y_B)$ with $\dot{\mathbf{x}} = (\dot{x}_A, \dot{y}_B)$, the initial state is $\mathbf{x}(0) = \mathbf{a}$, and the driving motion is $\dot{x}_A = 1$ for $t \in [0, 4]$. In free configurations **a**, **e**, $\dot{y}_B$ is zero. In contact configurations **b–d**, $\dot{y}_B$ is chosen to make $\dot{\mathbf{x}}$ tangent to the contact space normal, $\mathbf{n}$, which gives the equation $\mathbf{n} \cdot \dot{\mathbf{x}} = n_x\dot{x}_A + n_y\dot{y}_B = 0$. Substituting $\mathbf{n} = (-1, 1)$ and $\dot{x}_A = 1$ yields $\dot{y}_B = 1$.

5.2.1 Simulation Algorithm

Standard simulation algorithms require continuous simulation formulas. However, kinematic equations are discontinuous at changes in contact. The discontinuities are handled by splitting the simulation into intervals of fixed contact and simulating each interval separately. In the example of

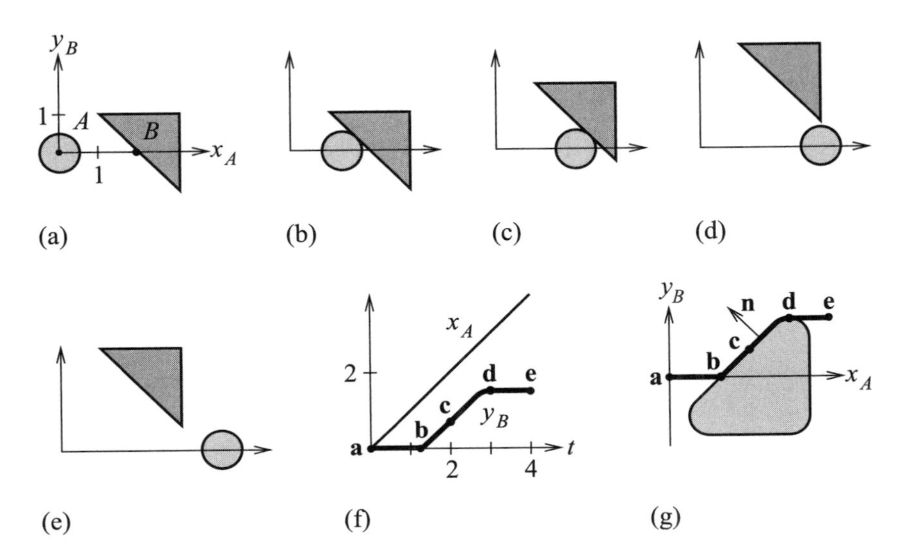

Figure 5.10
(a–e) Kinematic simulation of a translation pair, (f) time plots, (g) motion path in configurations **a–e** configuration space.

Input: mechanism, x(0), d, t_e, h, ϵ.
1. Compute pair partitions.
2. Set $i = 0$ and $t_0 = 0$.
3. While $t_i < t_e$
   a. Compute t_{i+1} and $x(t_{i+1})$.
   b. Set $i = i + 1$.

Output: $x\ (t_0), \ldots, x(t_e)$.

Figure 5.11
Kinematic simulation algorithm.

figure 4.3, the fixed contact intervals are from **a** to **b**, from **b** to **d**, and from **d** to **e**.

Figure 5.11 summarizes the kinematic simulation algorithm. The input is a mechanism, an initial configuration $\mathbf{x}(0)$, a driving motion d, a stop time t_e, a time step h, and an accuracy ϵ. The output is a sequence of configurations $\mathbf{x}(t_0), \ldots, \mathbf{x}(t_e)$. Configuration $\mathbf{x}(t_{i+1})$ is computed by integrating the equations $\dot{\mathbf{x}} = \mathbf{f}(\mathbf{x}, t)$ from t_i to t_{i+1} with accuracy ϵ. When the contacts of the parts are fixed on $[t_i, t_i + h]$, t_{i+1} equals $t_i + h$; otherwise, t_{i+1} is the contact change time.

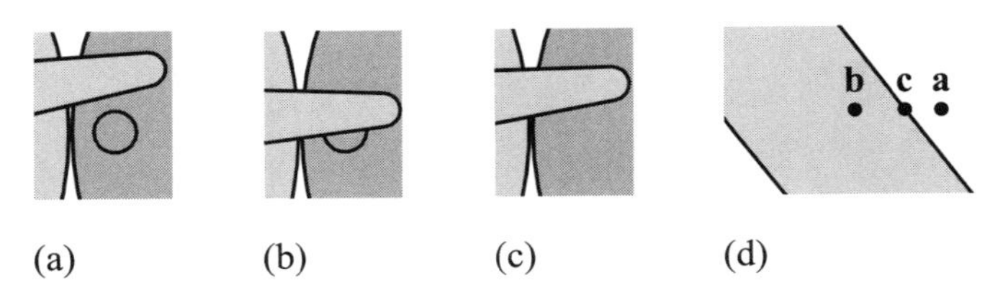

Figure 5.12
Collision: (a) free configuration **a**, (b) overlap configuration **b**, (c) contact configuration **c**, (d) configuration space partition.

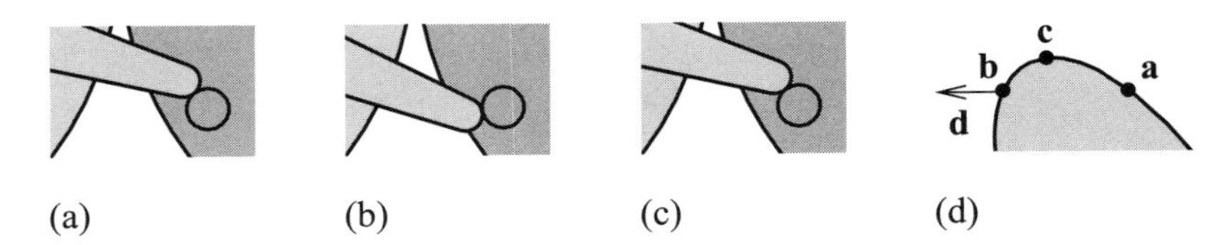

Figure 5.13
Contact breaks: (a) contact configuration **a**, (b) free velocity configuration **b**, (c) free configuration **c**, (d) configuration space partition.

The inputs to the integrator are the pairs' partitions, $\mathbf{x}(t_i)$, d, h, ϵ, and the velocity computation module described later. The integrator first computes $\mathbf{x}(t_i + h)$ under the assumption that the contacts are fixed on $[t_i, t_i + h]$. It uses a standard ordinary differential equation solver that invokes the velocity module multiple times to ensure ϵ accuracy. It then tests for contact changes. Two parts make contact when their configuration enters blocked space. A contact breaks when the velocity points into free space. Figures 5.12 and 5.13 illustrate this using the driver-indexer pair of the indexing mechanism from section 4.6. When a change is detected, the integrator finds the change configuration by a bisection search on $[t_i, t_i + h]$.

5.2.2 Computation of Velocity for Fixed-Axis Mechanisms

The input to the velocity computation algorithm for fixed-axis mechanisms is a configuration and a driving velocity. The algorithm assumes one driving part, one contact per pair, and no cycles in the topology graph of the mechanism. When these assumptions are violated, the simulation is overconstrained, so the mechanism blocks.

We first discuss a pair with $\mathbf{x} = (x_A, x_B)$ and a driving velocity $\dot{x}_A$. Define $\mathbf{v} = (\dot{x}_A, 0)$. When $\mathbf{x}$ is free, $\dot{x}_B = 0$ and $\dot{\mathbf{x}} = \mathbf{v}$. Suppose contact occurs with contact normal $\mathbf{n}$. If $\mathbf{n} \cdot \mathbf{v} \geq 0$, $\mathbf{v}$ points into free space or is tangential to

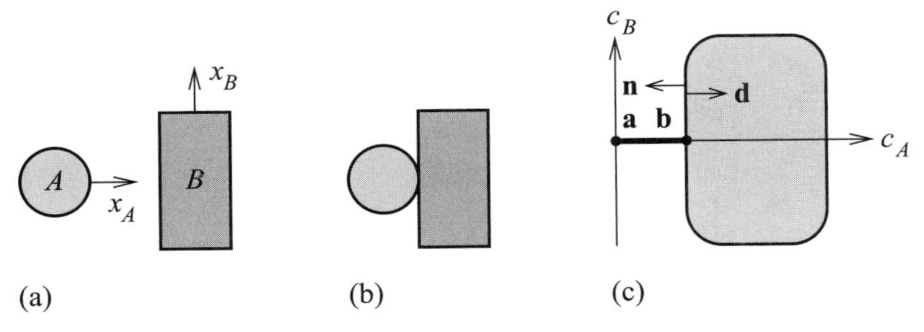

Figure 5.14
Blocked simulation: (a–b) configurations **a–b**, (c) motion path.

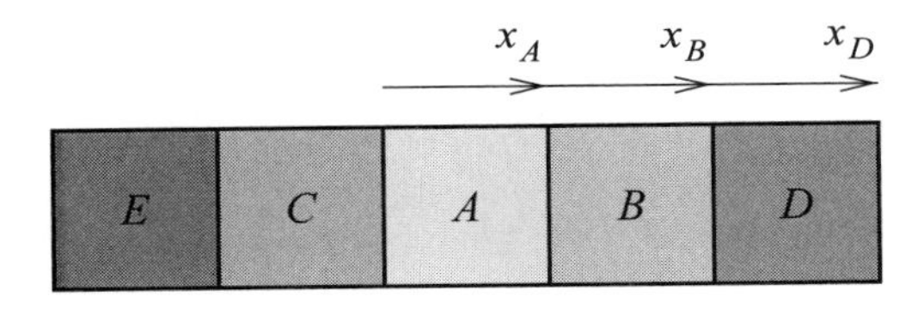

Figure 5.15
Computation of a mechanism's velocity.

contact space, A does not drive B and $\dot{\mathbf{x}} = \mathbf{v}$. The translation pair exhibits this behavior in configuration **d** in figure 5.10d, g. If $\mathbf{n} \cdot \mathbf{v} < 0$, $\mathbf{v}$ causes overlap of parts and a tangential $\dot{\mathbf{x}}$ is obtained by solving $\mathbf{n} \cdot \dot{\mathbf{x}} = n_x \dot{x}_A + n_y \dot{x}_B = 0$ for $\dot{x}_B$. There is no solution when $n_y = 0$, so we set $\dot{\mathbf{x}} = (0, 0)$. The simulation blocks because the driving velocity $\mathbf{v}$ is parallel to the contact normal $\mathbf{n}$ (figure 5.14).

The computation algorithm for a mechanism's velocity is as follows. The topology graph for the input configuration is traversed starting at the driving part. When part B is traversed from A, the pair algorithm is invoked with A as the driving part. If B is assigned a velocity, its children are traversed. In figure 5.15, part A is assigned a driving velocity $\dot{x}_A$, B is traversed from A and is assigned $\dot{x}_B = \dot{x}_A$, D is traversed from B and is assigned $\dot{x}_D = \dot{x}_B$, C is traversed from A without being assigned a velocity, and so E is not traversed.

Figure 5.16 shows a kinematic simulation of the indexing mechanism. The driving velocity is $\dot{\theta}_D = -1$ hertz. In figure 5.16a, the driver assigns $\dot{\theta}_I$ to the indexer, which assigns $\dot{\theta}_P$ to the pawl. The indexer time plot (figure 5.16d) shows that $\dot{\theta}_I$ is positive. The driver-indexer partition (figure 5.16f) shows that this is because $\dot{\theta}_D$ and the contact curve slope are negative. The

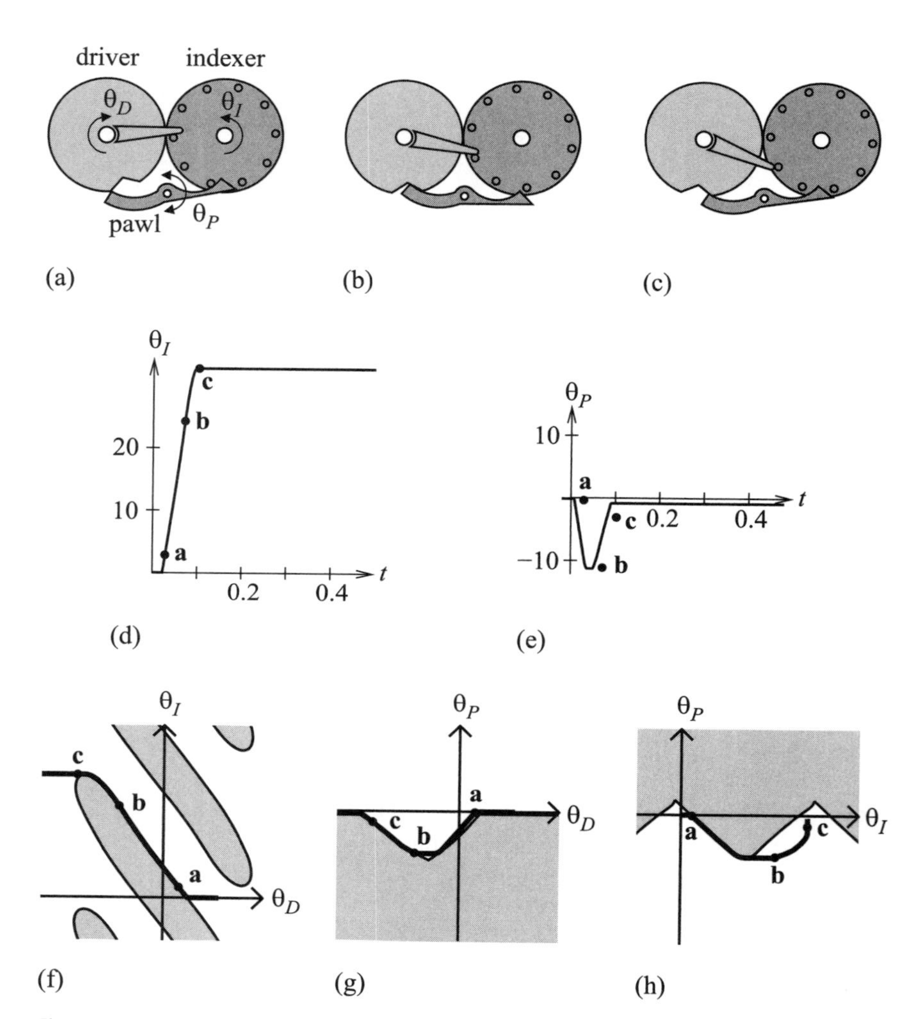

Figure 5.16
Kinematic simulation of an indexing mechanism: (a–c) configurations **a–c**, (d–e) time plots, (f–h) motion paths in configuration space.

pawl time plot (figure 5.16e) shows that $\dot{\theta}_P$ is negative and the indexer-pawl partition (figure 5.16h) explains why. In figure 5.16b, the driver assigns velocities to the indexer and to the pawl. In figure 5.16c, the driver assigns a velocity to the pawl and the indexer is stationary.

5.2.3 Computation of Velocity for General Planar Mechanisms

The algorithm for general planar mechanisms allows multiple driving parts and general topology graphs for a mechanism. The driving velocity is expressed as $\mathbf{d} = (\dot{x}_1, \dot{y}_1, \dot{\theta}_1, \ldots, \dot{x}_m, \dot{y}_m, \dot{\theta}_m)$ with m the number of moving parts. The driving parts contribute non-zero elements to $\mathbf{d}$ and the other elements are zero.

A contact between parts i and j is analyzed in the i-j configuration space with coordinates $(x_i, y_i, \theta_i, x_j, y_j, \theta_j)$. The driving velocity of the pair is $\mathbf{d}_{ij} = (\dot{x}_i, \dot{y}_i, \dot{\theta}_i, \dot{x}_j, \dot{y}_j, \dot{\theta}_j)$ and the contact normal is $\mathbf{n}_{ij}$. If $\mathbf{n}_{ij} \cdot \mathbf{d}_{ij} > 0$, the contact breaks immediately and is ignored. Otherwise, overlap of parts is prevented by the velocity along the contact normal. The velocity is expressed as $k\mathbf{n}$ with $\mathbf{n}$ the injection of $\mathbf{n}_{ij}$ into the mechanism's coordinates. This means that $\mathbf{n}$ is a $3m$-vector whose only non-zero elements are the first three elements of $\mathbf{n}_{ij}$ in positions $3i$, $3i+1$, $3i+2$ and the last three elements in positions $3j$, $3j+1$, $3j+2$.

When part j is the frame, the j elements are omitted from $\mathbf{d}_{ij}$ and $\mathbf{n}_{ij}$. A lower pair is modeled with two contacts, as explained in chapter 3: two point and circle contacts for a revolute joint and two point and line contacts for a prismatic joint. These contacts always generate velocities because they cannot break.

The mechanism's velocity is $\dot{\mathbf{x}} = \mathbf{d} + \sum_{i=1}^{c} k_i \mathbf{n}_i$ with c the number of contacts. We compute k_i from the c equations $\dot{\mathbf{x}} \cdot \mathbf{n}_i = 0$, which state that $\dot{\mathbf{x}}$ is tangent to the contact normals. The resulting symmetric linear system

$$\begin{bmatrix} \mathbf{n}_1 \cdot \mathbf{n}_1 & \cdots & \mathbf{n}_1 \cdot \mathbf{n}_c \\ \vdots & \ddots & \vdots \\ \mathbf{n}_c \cdot \mathbf{n}_1 & \cdots & \mathbf{n}_c \cdot \mathbf{n}_c \end{bmatrix} \begin{bmatrix} k_1 \\ \vdots \\ k_c \end{bmatrix} = \begin{bmatrix} -\mathbf{n}_1 \cdot \mathbf{d} \\ \vdots \\ -\mathbf{n}_d \cdot \mathbf{d} \end{bmatrix}$$

is solved with a linear equation solver. When the equations are singular, the simulation is overconstrained, so we assign the parts zero velocity.

We illustrate the algorithm using the constant-breadth cam pair from section 4.4 (figure 5.17). The cam is part 1 and the follower is part 2. The driving velocity is $\dot{\theta}_1 = 1$ hertz, so $\mathbf{d} = (0, 0, 1, 0, 0, 0)$. The revolute joint that connects the cam to the frame is modeled with contacts between cam points $\mathbf{a}^1 = (1, 0)$ and $\mathbf{b}^1 = (0, 1)$ and the frame circle, s, with a center

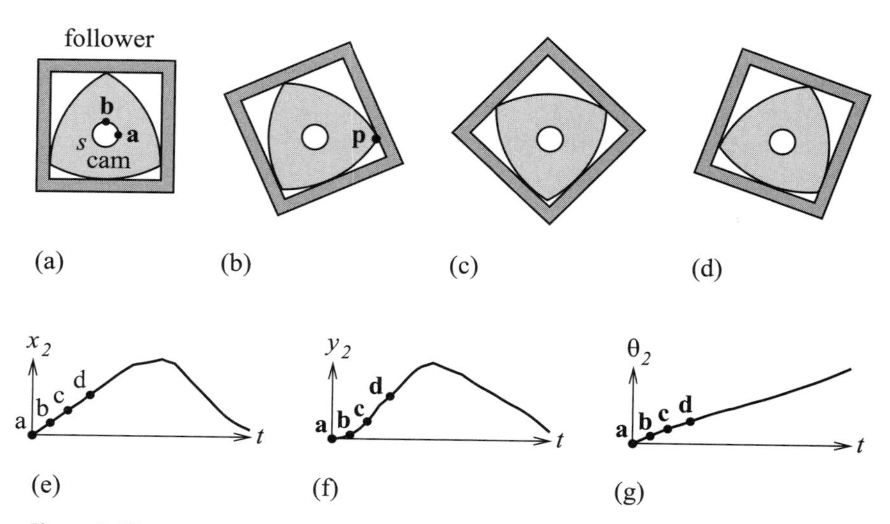

Figure 5.17
Kinematic simulation of a cam pair: (a–d) configurations **a–d**, (e–g) time plots.

$\mathbf{o}_b = (0, 0)$ and radius $r_b = 1$. The contact equations are $\|\mathbf{a}\| = 1$ and $\|\mathbf{b}\| = 1$, which yields

$$x_1^2 + y_1^2 + 2x_1 \cos\theta_1 + 2y_1 \sin\theta_1 = 0$$

$$x_1^2 + y_1^2 - 2x_1 \sin\theta_1 + 2y_1 \cos\theta_1 = 0.$$

The contact normals (after division by 2) are

$$\mathbf{n}_1 = (x_1 + \cos\theta_1, y_1 + \sin\theta_1, y_1 \cos\theta_1 - x_1 \sin\theta_1, 0, 0, 0)$$

$$\mathbf{n}_2 = (x_1 - \sin\theta_1, y_1 + \cos\theta_1, -y_1 \sin\theta_1 - x_1 \cos\theta_1, 0, 0, 0).$$

Figure 5.17a shows the configuration $(0, 0, 0, 0, 0, 0)$ where the cam-follower pair is free, so the mechanism's velocity is $\dot{\mathbf{x}} = \mathbf{d} + k_1\mathbf{n}_1 + k_2\mathbf{n}_2$. The linear equations are $k_1 = 0$ and $k_2 = 0$, since $\mathbf{n}_1 = (1, 0, 0, 0, 0, 0)$ and $\mathbf{n}_2 = (0, 1, 0, 0, 0, 0)$, so $\dot{\mathbf{x}} = \mathbf{d}$. Figure 5.17b shows the configuration $(0, 0, 29.81, 0.18, 0.04, 23.94)$ where the pair is in contact, so the velocity is $\dot{\mathbf{x}} = \mathbf{d} + k_1\mathbf{n}_1 + k_2\mathbf{n}_2 + k_3\mathbf{n}_3$. The contact is between cam point $\mathbf{p}^1 = (4.3, -2.5)$ and the follower line with normal $\mathbf{n}^2 = (-1, 0)$ and distance d from the origin. The contact equation is $\mathbf{n} \cdot \mathbf{f} = d$ with $\mathbf{f} = \mathbf{p} - \mathbf{t}_2$, which yields

$$\mathbf{n}_3 = (-\cos\theta_2, -\sin\theta_2, k, \cos\theta_2, \sin\theta_2, f_x \sin\theta_2 - f_y \cos\theta_2)$$

$$k = (p_x \cos\theta_2 + p_y \sin\theta_2) \sin\theta_1 + (p_y \cos\theta_2 - p_x \sin\theta_2) \cos\theta_1.$$

The contact normals are $\mathbf{n}_1 = (0.87, 0.5, 0, 0, 0, 0)$, $\mathbf{n}_2 = (-0.5, 0.87, 0, 0, 0, 0)$, and $\mathbf{n}_3 = (-0.91, -0.41, -2.04, 0.91, 0.41, 2.01)$. The linear equations are

$$\begin{bmatrix} 1 & 0 & -0.99 \\ 0 & 1 & 0.1 \\ -0.99 & 0.1 & 10.2 \end{bmatrix} \begin{bmatrix} k_1 \\ k_2 \\ k_3 \end{bmatrix} = \begin{bmatrix} 0 \\ 0 \\ 2.04 \end{bmatrix}$$

and $\dot{\mathbf{x}} = (0, 0, 0.55, 0.2, 0.09, 0.45)$. The driver rotates, the follower rotates and translates, and the sum of the angular velocities equals the driving angular velocity. Unlike the fixed-axis algorithm, the driving velocity is modified in the same way as the other velocities of parts.

5.3 Dynamical Simulation

Kinematic simulation presupposes that the driving velocities are known. This is a reasonable assumption when every degree of freedom in the mechanism is under accurate control. For example, the driver's orientation in our driver-indexer pairs can be controlled with a torque controller on a torsional motor. More generally, each part of a mechanism moves according to the Newtonian equations

$$m\ddot{x} = f_x + c_x$$

$$m\ddot{y} = f_y + c_y$$

$$I\ddot{\theta} = \tau + c_\theta$$

with m the mass of the part, $\ddot{x}$ and $\ddot{y}$ the components of its linear acceleration, f_x and f_y the external force acting on the part, c_x and c_y the contact force, I the moment of inertia, $\ddot{\theta}$ the angular acceleration, τ the external torque, and c_θ the contact torque.

The rigid part model contradicts the Newtonian equations when parts collide. The parts must change velocity instantaneously to prevent overlap, but the velocity is continuous and only acceleration can change instantaneously. The contradiction is resolved by employing a separate collision model. When the parts are smooth, the postcollision velocities are derivable from the conservation of momentum. Continuum mechanics provides a more general model in which parts deform during collision, then return to their original shapes.

Numerical solution of the extended Newtonian equations is called dynamical simulation. It provides designers of mechanisms with information that goes beyond the realm of kinematics. The most robust designs

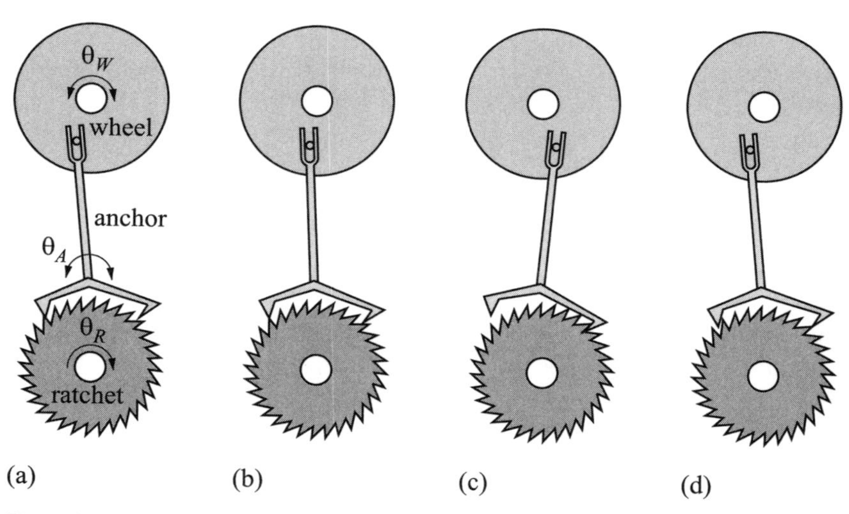

Figure 5.18
Dynamical simulation of escapement: (a–d) configurations **a–d**.

have a single kinematic function, meaning a single motion path in configuration space, so dynamical simulation differs from kinematic simulation solely in the rate at which this path is traversed. When the function depends on dynamics, dynamical simulation identifies the actual function of a specific design.

Example We illustrate dynamical simulation for a fixed-axis escapement composed of a wheel, an anchor, and a ratchet (figure 5.18). The kinematic function is to rotate the ratchet clockwise by two teeth per second. A spiral spring oscillates the wheel twice per second. The wheel pin engages the anchor fork and oscillates the anchor. A weight applies a constant clockwise torque to the ratchet. The cycle begins with the wheel pin at its leftmost point (figure 5.18a). The left anchor pallet engages a ratchet tooth and blocks its rotation. As the wheel rotates counterclockwise, the anchor rotates clockwise. When the wheel pin is at its lowest point, the ratchet disengages the left anchor pallet (figure 5.18b), rotates clockwise by one tooth, and engages the right anchor pallet (figure 5.18c). As the wheel rotates clockwise, the right anchor pallet disengages (figure 5.18d), the left pallet engages, and the cycle repeats.

Dynamical simulation verifies the kinematic function for one design instance. The moments of inertia are 1 $\mathrm{N \cdot cm^2}$ (newton for square centi-

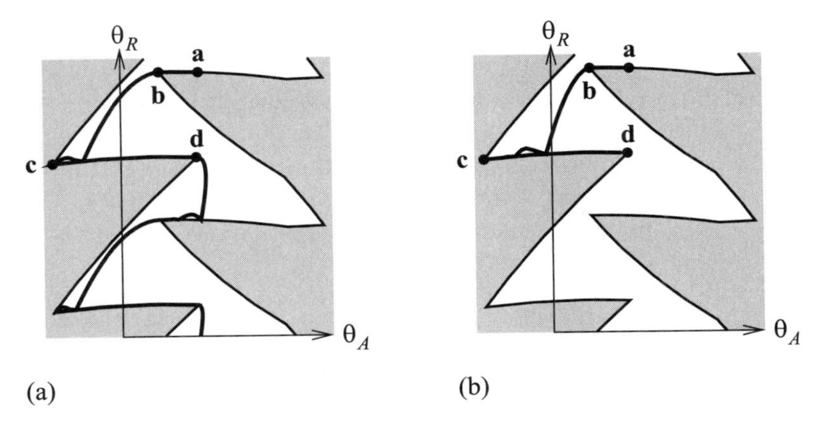

Figure 5.19
Anchor-ratchet motion paths: (a) correct, (b) failure.

meter) for the balance, 1 $\mathrm{N} \cdot \mathrm{cm}^2$ for the anchor, and 5 $\mathrm{N} \cdot \mathrm{cm}^2$ for the escapement. Friction is neglected. A spring coefficient of π^2 produces a natural spring period of two cycles per second. Simulating the mechanism with a range of driving torques reveals that it works with 20 $\mathrm{N} \cdot \mathrm{cm}$, but fails with 30 $\mathrm{N} \cdot \mathrm{cm}$. Figure 5.19 shows the two anchor-ratchet motion paths. The bad path (figure 5.19b) starts correctly but fails to clear the horizontal contact curve at configuration **d**, reverses direction, and blocks in configuration **c**.

Algorithm The main steps in dynamical simulation are computation of contact force, impact handling, and integration. Contacts between smooth parts generate forces along the contact normals that are computed by solving linear equations, just as velocities are computed in kinematic simulation. Contacts between rough parts generate tangential, frictional forces that are harder to compute. Impact handling updates the velocities of colliding parts according to the conservation of momentum, continuum mechanics, or variants thereof. The integration algorithm is the same as in kinematic simulation except that it uses the Newtonian equations and invokes the impact handler on collisions.

Dynamical simulation applies to spatial parts. Each part has three rotational and three translational equations. Collisions are detected by testing the parts for overlap since partitions of configuration space are impractical. Efficient detection of collisions is widely studied in mechanical engineering and robotics.

Simulation with Configuration Space Partitions The key advantage of using configuration space partition for simulation is that it provides a model for contact of parts. This model is an effective alternative to precomputed pairwise contacts and collision detection. Partition of the configuration space encodes the contact conditions and relations that must be incorporated in the equations for the motion of parts. It thus allows the efficient and robust simulation of mechanisms with complex contact geometry and complex changing sequences of contact.

5.4 Notes

Analysis of mechanisms plays a central role in mechanical engineering. The main analysis tools are kinematic and dynamical simulation, which have been the subject of extensive research in the past two decades. A variety of commercial and academic software packages are currently available, including ADAMS, SIMPACK, DADS, and SD/FAST, to name a few. The simulators handle multibody systems composed of parts connected by permanent joints. For textbooks on simulating mechanisms, see de Jalon and Bayo [32], Haug [28], Schiehlen [77], and Nikravesh [59].

We have studied analysis in the configuration space paradigm. We show how configuration space partition regions and their adjacencies correspond to kinematic function for fixed-axis mechanisms [36]. Based on this representation, we developed a symbolic language to describe kinematic function [39], that can be used to compare and classify mechanisms [37, 38]. We show how to compute and interpret the configuration space partition for fixed-axes mechanisms [40]. See also related work on qualitative kinematics by Faltings [23, 24].

Appendix B describes the HIPAIR kinematic simulation format and visualization control.

6 Tolerancing

This chapter addresses tolerancing, which is the task of transforming a nominal design into one that can be manufactured. Every manufacturing process is subject to variation in the shape and configuration of parts that causes kinematic variation. The manufacturing cost generally increases as the variation decreases. The designer needs to select a manufacturing process and to ensure that it produces mechanisms that function as intended despite kinematic variation.

Variation in mechanisms is represented with tolerances. We use parametric tolerances because they are general and conform with tolerancing practice. The variation in manufacturing is modeled with error intervals around the nominal parameter values of the mechanism's parametric model. The model is analyzed to estimate the kinematic variation. When the variation is excessive, the mechanism is redesigned, as discussed in the next chapter, or a more accurate manufacturing process is selected.

We present algorithms that derive the worst-case kinematic variation of mechanisms with parametric tolerances. We describe the tolerance specifications in section 6.1, kinematic variation in section 6.2, and algorithms for kinematic pairs and mechanisms in sections 6.3 and 6.4.

6.1 Specifications of Parametric Tolerance

We saw in section 2.1 that a parametric model specifies parts in terms of parameters $\mathbf{p} = (p_1, \ldots, p_n)$. A boundary segment is specified with an implicit curve, $f(x, y, \mathbf{p}) = 0$, that depends on $\mathbf{p}$. The model contains parameters for all the part boundaries and motion axes of the mechanism. We specify nominal parameter values, $\bar{\mathbf{p}} = (\bar{p}_1, \ldots, \bar{p}_n)$, and lower and upper parameter bounds, $\mathbf{l} = (l_1, \ldots, l_n)$ and $\mathbf{u} = (u_1, \ldots, u_n)$. The tolerance intervals, $\bar{p}_i - l_i \leq p_i \leq \bar{p}_i + u_i$, model the variation of the manufacturing process.

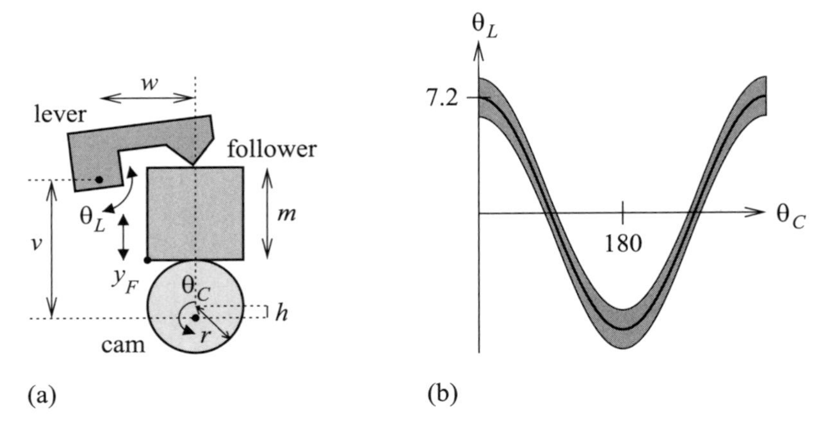

Figure 6.1
(a) Lever mechanism and (b) kinematic variation.

Figure 6.1a shows a parametric model of a lever mechanism composed of an eccentric rotating cam, a translating follower, and a rotating lever. The parameters are $\mathbf{p} = (r, h, m, w, v)$ with r and h the cam's radius and eccentricity, m the follower's axis height, and w and v the axis positions of the horizontal and vertical levers. The tolerance specifications are $\overline{\mathbf{p}} = (1, 0.25, 2, 2, 3)$ and $l_i, u_i = 0.01$.

The cross-product of the tolerance intervals, a hyperbox in the parameter space, is the tolerance space. Each point in the tolerance space is an instance of the nominal design. We assume that the shapes of the parts have the nominal topology in every instance; incident segments have equal endpoints and nonincident segments are disjoint.

6.2 Tolerance Analysis

The task of worst-case tolerance analysis is to compute the maximal variation from the nominal kinematic function over the tolerance space. We illustrate this using the lever mechanism, which has fixed contacts, then define kinematic variation for general mechanisms.

The driving motion of the lever mechanism is rotation of the cam. The cam raises and lowers the follower, which raises and lowers the lever. The kinematic function is the relation between the cam angle, θ_C, and the lever angle, θ_L. Figure 6.1b shows the nominal function (thick curve) and its variation (shaded area). The variation is maximal when $\theta_C = 0°$ and $\theta_C = 180°$ and is minimal when $\theta_C = 90°$ and $\theta_C = 270°$. The boundary curves of the variation region represent the extremal values of θ_L at each θ_C value.

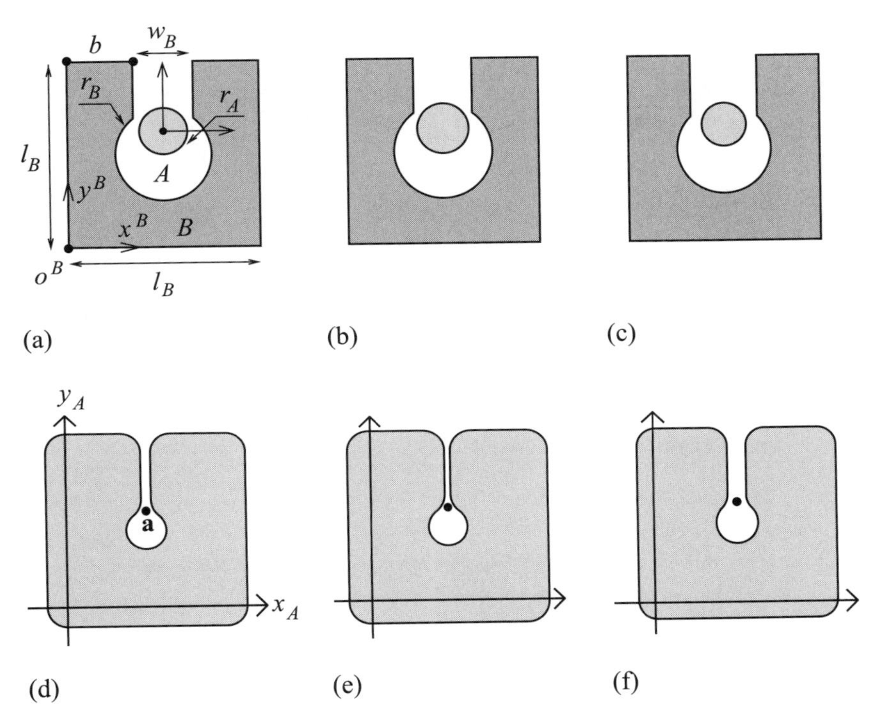

Figure 6.2
Circle-box pair: (a–c) nominal shapes and two valid instances in configuration $\mathbf{a} = (4, 5)$, (d–f) nominal and instance partitions of configuration space.

The kinematic function of a mechanism is its partition of configuration space. Each instance in a tolerance space has its own partition, owing to its the shapes and motion axes of its parts. The variation in the partition represents the kinematic variation of the mechanism. The instance contact spaces form a band around the nominal contact space, called the contact zone, that comprises the configurations where some instance is in contact. The nominal free space minus the contact zone comprises the configurations that are free for every instance; likewise for contact space.

We illustrate contact zones with a pair (figure 6.2a) composed of a circle A of nominal radius $r_A = 1$ and a box B with a square outer profile of nominal length $l_B = 8$, a circular hole at its center of nominal radius $r_B = 2$, and a vertical slot of width $w_B = 2.5$. The clearance between circle A and the B slot is 0.5. The parameter bounds are 0.1 for r_A, r_B and 0.2 for l_B, w_B. Figure 6.2b shows an instance with $r_A = 1.1$, $l_B = 8.2$, $r_B = 2.1$, $w_B = 2.5$, and clearance 0.3. Figure 6.2c shows an instance with $r_A = 0.9$, $l_B = 7.8$, $r_B = 1.9$, $w_B = 2.7$, and clearance 0.9.

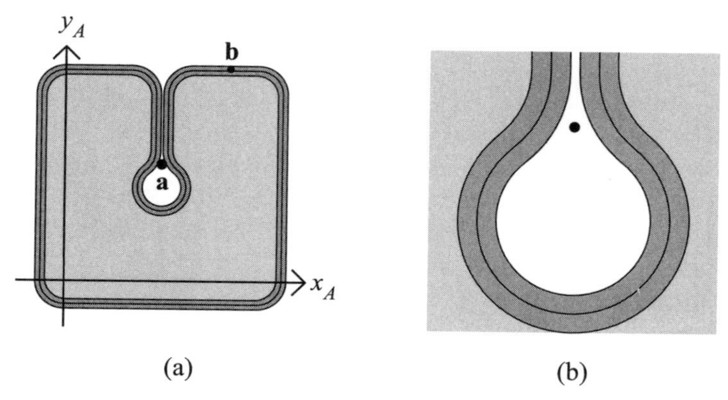

Figure 6.3
Circle-box pair contact zone (a) with detail (b).

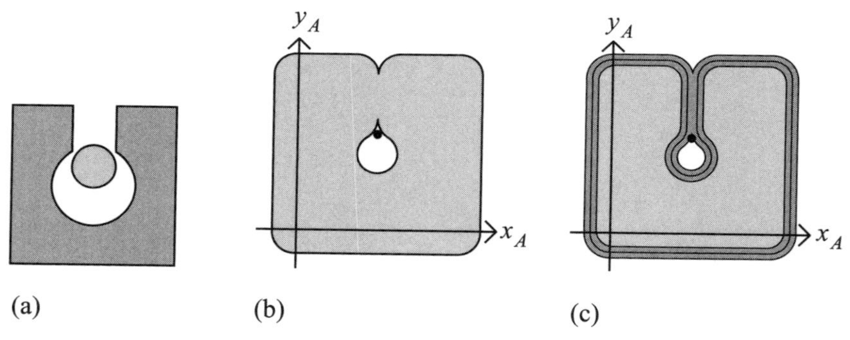

Figure 6.4
Circle-box pair with larger variation: (a) blocked instance, (b) configuration space partition, (c) contact zone.

The configuration space of the pair is (x_A, y_A) because A translates with $\theta_A = 0°$ and B is fixed. Figure 6.2d–f shows the nominal partition and the two variations. Figure 6.3 shows the contact zone: the dark gray band around the nominal contact space. Configuration $\mathbf{a} = (4, 5)$ is free for every instance, whereas $\mathbf{b} = (7, 9.1)$ is free for the nominal shape (figure 6.2a) and is blocked for the first valid instance (figure 6.2b).

A difference between the nominal partition topology and that of an instance indicates a qualitative change in kinematic function. In our example, the topology changes when the inner and outer contact zone boundaries intersect (figure 6.4c). This occurs when the bounds are 0.15 for r_A, r_B and 0.3 for l_B, w_B. There are instances in which A does not fit in the vertical slot, as illustrated in figure 6.4a–b with $r_A = 1.1$, $l_B = 8.2$, $r_B = 2.1$, $w_B = 2.2$. The narrow vertical channel disappears, indicating that the circle cannot exit the box.

In our example, the outer and inner contact zone boundaries correspond to the maximum and minimum material condition instances. This is not the case in general. Each boundary point is generated by an instance, but there need not exist one instance that generates the boundary. The contact zone is thus a conservative estimate of worst-case variation. In particular, a contact zone boundary intersection does not guarantee that any instance exhibits a qualitative change in kinematic function.

Disk Indexer Pair We illustrate a change in contact zone topology using the disk indexer pair from section 5.1.1 (figure 6.5). Figure 6.5a–c shows that a narrow contact zone has the topology of the nominal partition: the diagonal segments represent drive periods and the horizontal segments represent dwell periods. Figure 6.5d–f shows that a wider contact zone has a different topology in which the diagonal channels are replaced by lower and upper horns. The topology changes indicate that instances in the tolerance space might block, and the displayed instance shows that blocking can occur. Although the horizontal channels of the failure instance are blocked, the diagonal channels are open, so this aspect of the contact zone is conservative with regard to this instance.

Intermittent Gear Mechanism We illustrate general planar pair contact zones using the gear-follower pair of the intermittent gear mechanism from section 5.1.3 (figure 6.6). The parameters that specify the gear teeth and the follower's pawls are assigned tolerances of ± 0.01 mm. The contact zone has the nominal topology at gear orientations where a pawl drives a gear tooth, as shown in figure 6.6a and c. The topology is different when neither pawl is engaged (figure 6.6b): the nominal free region splits into two regions, which indicates that the follower blocks when the pawls engage opposing teeth.

6.3 Algorithms for Kinematic Pairs

Tolerance analysis of a kinematic pair consists of computing the contact zone. We employ boundary representations. A fixed-axis pair has a two-dimensional zone whose boundary consists of curve segments. A general planar pair has a three-dimensional zone whose boundary consists of surface patches. The boundaries are not expressible in closed form, except for very simple shapes of parts, so we compute approximate contact zones.

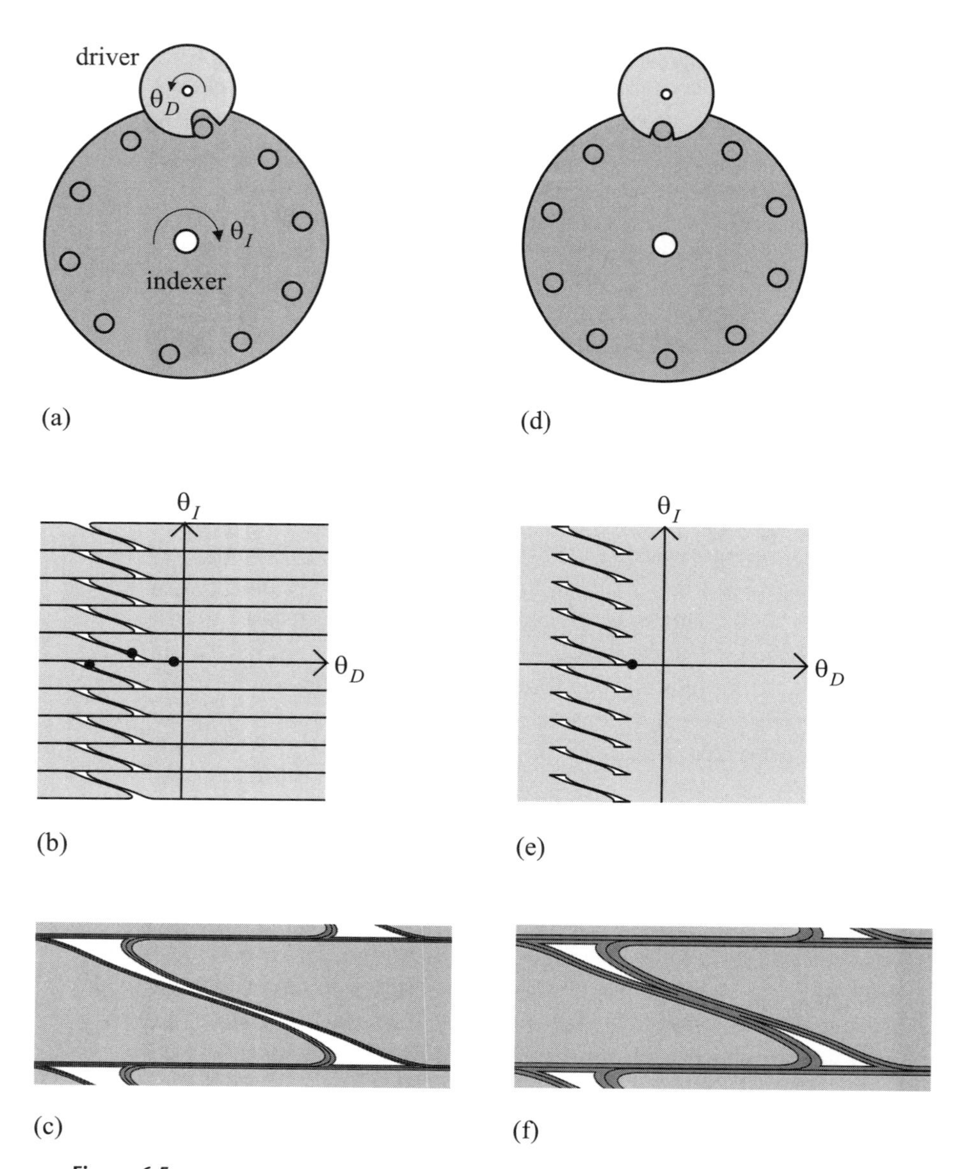

Figure 6.5
Disk indexer pair: (a) nominal design, (b) nominal partition, (c) narrow contact zone, (d) failure instance, (e) failure partition, (f) wider contact zone.

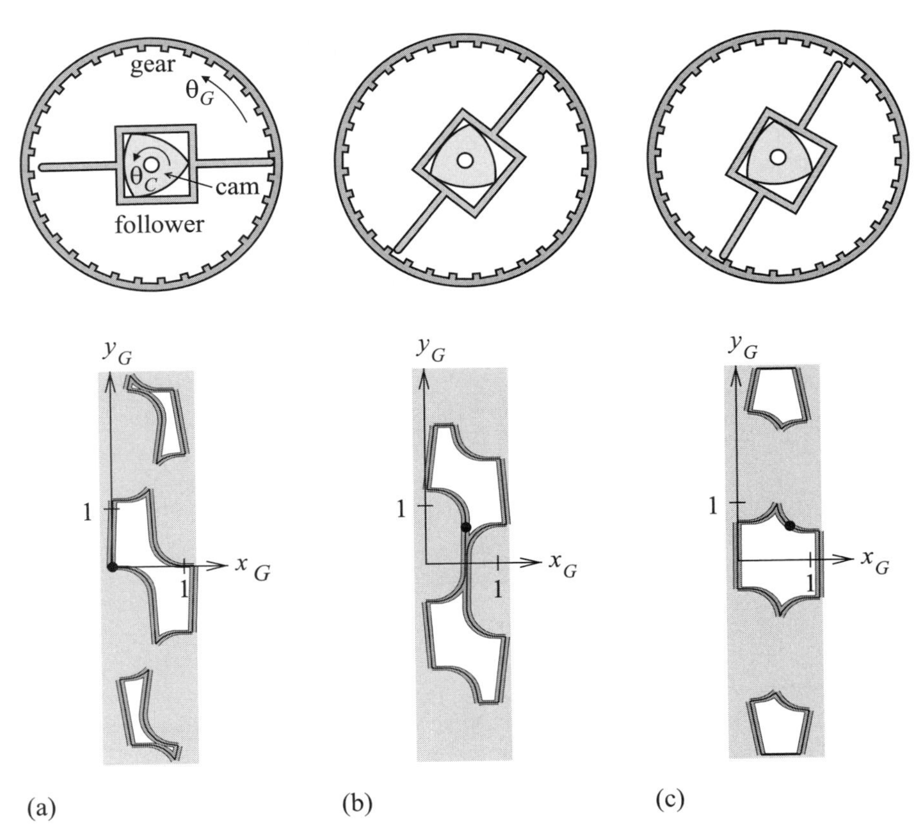

Figure 6.6
Gear-follower contact zone cross-sections: (a) $\theta_c = 0°$, (b) $\theta_c = 5°$, (c) $\theta_c = 9°$.

Figure 6.7 summarizes the algorithm. Steps 1 and 2 sample the contact space. For a fixed-axis pair, the contact curves are sampled as described in section 4.3.1. For a general planar pair, cross-sections are formed at a specified θ_A spacing and the resulting fixed-axis pairs are sampled. Step 3 computes the contact zone boundary points in the outward and inward normal directions for each sample of contact space. Step 4 forms the contact zone boundary from the boundary points. Figure 6.8 illustrates the algorithm for the figure 6.3 contact zone.

Step 3 The input is a configuration, $\mathbf{c}$, on a parametric contact curve or surface, $f(\mathbf{c}, \bar{\mathbf{p}}) = 0$. The outward normal at $\mathbf{c}$ is the gradient vector, $f_\mathbf{c}$. The contact zone's outer boundary point lies on the ray $\mathbf{c} + kf_\mathbf{c}$ and maximizes k

Input: parametric pair, parameter values and intervals, accuracy.
1. Compute partition.
2. Sample contact space to accuracy.
3. Compute contact zone boundary points.
4. Form contact zone boundary.
Output: contact zone.

Figure 6.7
Algorithm for the computation of the contact zone.

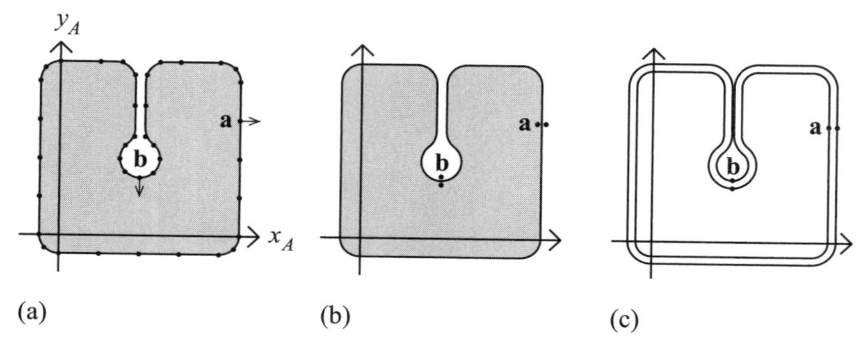

Figure 6.8
Contact zone computation: (a) contact space sampling, (b) contact zone boundary points at **a** and **b**, (c) contact zone boundaries.

under the constraints $f(\mathbf{c} + kf_\mathrm{c}, \bar{\mathbf{p}} + \mathbf{q}) = 0$ and $-l_i \leq q_i \leq u_i$. The inner boundary point minimizes k. We approximate f with its linear Taylor series

$$f(\mathbf{c} + kf_\mathrm{c}, \bar{\mathbf{p}} + \mathbf{q}) \approx f(\mathbf{c}, \bar{\mathbf{p}}) + kf_\mathrm{c} \cdot f_\mathrm{c} + \mathbf{q} \cdot f_\mathbf{p} = kf_\mathrm{c} \cdot f_\mathrm{c} + \mathbf{q} \cdot f_\mathbf{p}$$

and solve $f = 0$ for

$$k = -\frac{1}{f_\mathrm{c} \cdot f_\mathrm{c}} \mathbf{q} \cdot f_\mathbf{p} = -\frac{1}{f_\mathrm{c} \cdot f_\mathrm{c}} \sum_i q_i \frac{\partial f}{\partial p_i}.$$

This formula shows that k is maximal when $q_i = u_i$ for $\partial f/\partial p_i > 0$ and $q_i = -l_i$ otherwise. It is minimal when l_i and u_i are interchanged.

We illustrate the computation using the circle-box pair (figure 6.8b). The sole parameter is $p = r_a$ with $\bar{p} = 1$, $l = 0.2$, and $u = 0.2$. Configuration $\mathbf{a} = (9, 6)$ lies on the moving circle-fixed line contact curve $f(x_a, y_a, r_a) = x_a - r_a - 8 = 0$, using equation 3.7 with $\mathbf{n}_b = (1, 0)$, $\mathbf{o}_a^A = (0, 0)$, and $d_b = 8$. The gradient is $(\partial f/\partial x_a, \partial f/\partial y_a) = (1, 0)$ and $k = -q(\partial f/\partial r_a) = q$ with $-0.2 \leq q \leq 0.2$. The maximum, $k = 0.2$, occurs when q is largest and generates the contact zone's outer boundary point $(9.2, 6)$. The minimum, $k = -0.2$, occurs when q is smallest and generates the inner boundary point $(8.8, 6)$. Configuration $\mathbf{b} = (4, 3)$ lies on the moving circle-fixed circle contact curve $(2 - r_a)^2 - (x_a - 4)^2 - (y_a - 4)^2 = 0$, using equation 3.8 with $\mathbf{o}_a^A = (0, 0)$, $\mathbf{o}_b^B = (4, 4)$, and $r_b = 2$. The gradient is $2(4 - x_a, 4 - y_a) = (0, -2)$ and $k = 0.2q$. The maximum and minimum generate the outer and inner boundary points $(4, 2.8)$ and $(4, 3.2)$.

The error that results from the linear Taylor series is quadratic in the tolerance interval widths. Usually this error is negligible because the tolerance intervals are narrow, but there are cases where the error is significant. We can derive the true maximum and minimum via constrained nonlinear optimization, which takes longer than linear optimization, but is still practical.

Step 4 In a fixed-axis pair, each free region boundary yields an outer and an inner sequence of contact zone boundary points. We link the points in each sequence to obtain curves (figure 6.8c). These curves partition the configuration space into faces. The faces that contain the contact space are the contact zone. When the tolerance intervals are narrow, the contact zone is a narrow band around the contact curves (figure 6.3). When the intervals are wide enough, the curves intersect and the contact zone topology is more complicated (figure 6.4). In a general planar pair, the approximate contact zone boundaries are triangular meshes that generate a spatial partition.

6.4 Algorithms for Mechanisms

The computation algorithm for a contact zone cannot be used for mechanisms because we cannot in general construct mechanism partitions. If we could, the algorithm would extend directly. Instead we compute error bounds at sample configurations on the nominal path for a given driving motion, which we compute by kinematic simulation (section 5.2). A configuration, $\mathbf{c}$, lies on a set of contact surfaces, $f^j(\mathbf{c} + \mathbf{d}, \bar{\mathbf{p}} + \mathbf{q}) = 0$, that linearize to $f_{\mathbf{c}}^j \cdot \mathbf{d} + f_{\mathbf{p}}^j \cdot \mathbf{q} = 0$. We seek a maximal $\mathbf{d}$ subject to the linearized f^j constraints and $-l_i \leq q_i \leq u_i$. The $\mathbf{d}$ metric is task specific. A common criterion is that an output configuration variable be maximized with the driving variables fixed at their nominal values.

We illustrate using a lever mechanism (figure 6.1). The configuration space coordinates are $(\theta_C, y_F, \theta_L)$. The tolerance parameter vector is $\mathbf{p} = (r, h, m, w, v)$ with $\bar{\mathbf{p}} = (1, 0.25, 2, 2, 3)$ and with $l_i, u_i = 0.01$. The cam-follower contact function is $y_F - p_1 - p_2 \cos\theta_C = 0$ and the follower-lever contact function is $p_4 \sin\theta_L + p_5 - p_3 - y_F = 0$. We compute the variation in θ_L. Eliminating y_F yields $p_4 \sin\theta_L + p_5 - p_3 - p_1 - p_2 \cos\theta_C = 0$ and linearizing with θ_C fixed yields

$$d_3 p_4 \cos\theta_L = -q_5 + q_3 + q_1 + q_2 \cos\theta_C - q_4 \sin\theta_L.$$

The d_3 maximum is $(0.03 + 0.01|\cos\theta_C| + 0.01|\sin\theta_L|)/\cos\theta_L$ and the minimum is its negative. The maximum variation, $d_3 = 2.4°$, occurs at $\theta_C = 0°$ and $\theta_C = 180°$ where θ_L is extremal. The minimum variation, $d_3 = 1.7°$, occurs at $\theta_C = 90°$ and $\theta_C = 270°$ where $\theta_L = 0$.

6.5 Notes

Tolerancing of mechanisms is of great practical and economic importance. Worst-case tolerancing sets the stage for statistical tolerancing in which manufacturing variation is modeled with probability distributions and kinematic variation is treated as a random variable. A manufacturing process is accepted based on the probability that its output will function correctly or on its expected cost.

Current tolerancing principles and standards are described by the American Society of Mechanical Engineers [1, 2]. Popular tolerance specifications are parametric and geometric [66, 80]. A variety of methods for worst-case and statistical tolerancing of mechanisms have been proposed [9]. For a survey of commercial computer-aided tolerancing (CAT) packages, see Prisco [63]. Tolerancing for assembly, whose task is to ensure that the manufactured parts will assemble, is a closely related topic [82].

The most common kinematic tolerance analysis is tolerance chain or stack-up analysis for static and small displacements. It consists of identifying critical parameters, such as a gap, clearance, or play; building a one-dimensional tolerance chain based on configurations and contacts of parts; and determining the parameters' variability range [8, 17, 81]. Kinematic variations for large displacements are studied in Chase et al. [13].

We have developed algorithms for kinematic tolerancing in the configuration space paradigm [41, 72, 73, 75]. We automate computation of kinematic variation with contact changes. The algorithms use the linear Taylor series approximation described earlier. In later work [44] we developed an exact algorithm based on constrained optimization. The linear approximation is normally accurate, but is sometimes very inaccurate at critical configurations.

Appendix B describes HIPAIR computation of contact zones. HIPAIR handles fixed-axis planar pairs with uniform offset tolerances.

7 Synthesis

This chapter studies the design task of kinematic synthesis. The goal is to design a mechanism that realizes a kinematic function. Synthesis is the inverse process of analysis. It is inherently more difficult because many mechanisms can realize the same kinematic function. We focus on parameter synthesis: given a parametric model of a mechanism, a kinematic function, and its allowed variability, the goal is to find nominal parameter values and tolerance intervals that realize the kinematic function with the allowed variability.

In section 7.1 we review the kinematic design cycle before focusing on parameter synthesis. In section 7.2 we present a parameter synthesis algorithm that finds parameter values that realize a kinematic function. In section 7.3, we describe a parameter optimization algorithm that adjusts parameter values to ensure correct kinematic function for a given tolerance space. We discuss mechanisms composed of fixed-axis higher pairs.

7.1 Kinematic Design Cycle

We saw in chapter 1 that kinematic design is an iterative process with conceptual design, parametric design, analysis, synthesis, and tolerancing steps. Conceptual design consists of selecting a design concept that captures the desired kinematic function. Parametric design consists of building a parametric model of the mechanism that encodes the shapes and configurations of parts. Analysis derives the kinematic function of a mechanism from a specification of its parts' shapes and motion constraints. Synthesis is the inverse task of devising a mechanism that performs a specified function. Tolerancing derives the kinematic effect of manufacturing variation.

We illustrate this cycle with a driver-follower pair from an optical filter mechanism, which is discussed further in section 8.1. The design requirements are to cover and uncover a lens with a filter (figure 7.1a). Figure

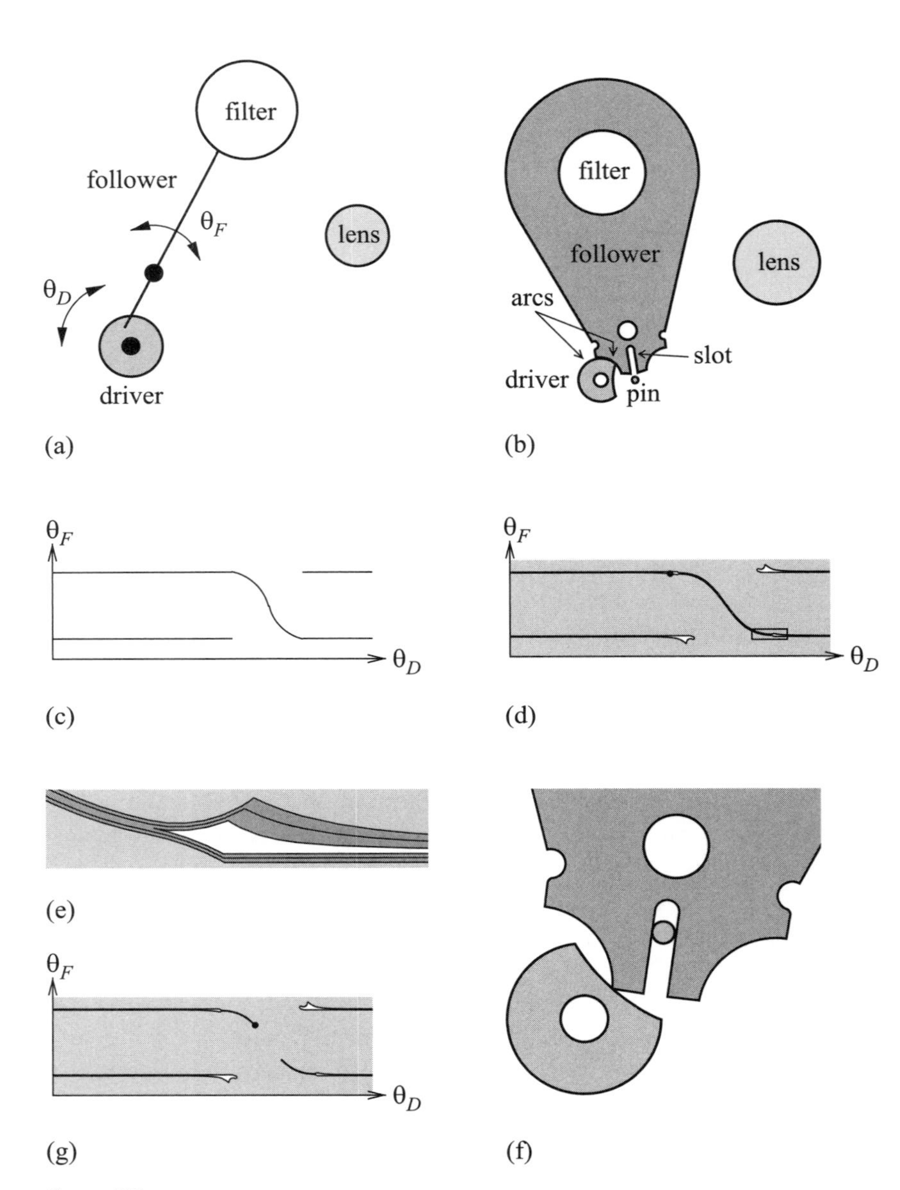

Figure 7.1
Kinematic design scenario: (a) design requirements, (b) nominal design, (c) intended kinematic function, (d) nominal partition, (e) contact zone detail, (f) failure instance, (g) failure partition.

7.1c shows the intended kinematic function. Counterclockwise rotation of the driver turns the follower clockwise, so the filter covers the lens, then locks it in place. Clockwise rotation unlocks the follower, rotates it until the filter uncovers the lens, and re-locks it.

During conceptual design, the designer chooses a Geneva pair. During parametric design, the geometry and rotation axes for the parts are chosen (figure 7.1b). The intended feature contacts are a pin and slot in the driving phase and concentric convex and concave arcs in the locking phase. The designer creates a parametric model and assigns parameter values and tolerance intervals. The nominal function is correct (figure 7.1d), but the contact zone (figure 7.1e) indicates possible blocking, owing to overlap between the upper and lower boundaries. Figure 7.1f–g shows a blocking instance and its partition. The designer can change the parameter values to remove the blocking, can tighten the tolerance intervals, or can change the design concept.

7.2 Parameter Synthesis

The input to our parameter synthesis algorithm is a parametric model of a mechanism with parameter intervals. The cross-product of the intervals is the design space. Each point in the design space is a design instance. We seek a design instance that realizes a specified kinematic function in an optimal manner.

A manual search of the design space is often impractical. The designer must examine many instances to ensure that a good design has not been overlooked. Each instance requires a time-consuming analysis. The search is especially difficult when the kinematic function is sensitive to small perturbations in the parameter values, as is common with higher pairs. Unlike tolerance spaces, design spaces are generally too large for linearization of the contact curves to have acceptable accuracy.

Our solution is an optimization search. We encode the design goals in objective functions that we minimize subject to the kinematic constraints. The design goals usually cannot be expressed with a single function because they are often incommensurate or conflicting. The entire design space cannot be searched because of its large dimension. These considerations lead us to an interactive paradigm in which the designer refines an initial design instance through a sequence of design changes.

Figure 7.2 summarizes the algorithm. The designer examines the configuration space partitions for design flaws, decides how to change the

Input: parametric model and initial parameter values.
1. Compute partitions.
2. Input design changes from partitions.
3. If no changes, return parameter values.
4. Iteratively minimize design changes function:
   a. Make optimization step.
   b. Prevent partition structure changes.
5. Update parameter values and go to step 2.

Output: final parameter values.

Figure 7.2
Algorithm for parameter synthesis.

partitions to remove the flaws, and inputs the changes. The algorithm translates the changes into an objective function and computes a local minimum starting from the current parameter values. The cycle repeats until the design is acceptable.

7.2.1 Design Change

A design change is specified with a dragger in configuration space: an arrow whose tail, $\mathbf{t}$, lies on a contact curve with equation $f(\mathbf{c}, \mathbf{p}) = 0$ and whose head, $\mathbf{h}$, is a configuration that should lie on that curve. The tail satisfies $f(\mathbf{t}, \mathbf{p}_i) = 0$ with $\mathbf{p}_i$ the initial design instance. The objective function is $|f(\mathbf{h}, \mathbf{p})|$ with parameters $\mathbf{p}$.

We illustrate draggers with the circle-box pair (figure 7.3). The initial design instance (figure 7.3a) and configuration space partition (figure 7.3b) are from section 6.2. The dragger (figure 7.3c) has a tail $\mathbf{t} = (3.75, 5.75)$ and head $\mathbf{h} = (3.9, 5.75)$. The contact is between A and the left side of the B slot. Its equation is $f(\mathbf{c}, \mathbf{p}) = x_A - r_A + 0.5w_B - 4 = 0$ with $\mathbf{c} = (x_A, y_A)$ and $\mathbf{p} = (r_A, w_B)$. The initial design instance is $\mathbf{p}_i = (1, 2.5)$. The objective function, $|f(\mathbf{h}, \mathbf{p})| = |3.9 - r_A + 0.5w_B - 4|$, is obtained by setting $\mathbf{c} = \mathbf{h}$ in the contact equation. Its value at $\mathbf{p} = \mathbf{p}_i$ is $|3.9 - 1 + 1.25 - 4| = 0.15$.

We minimize the objective function by iterative optimization starting from $\mathbf{p} = \mathbf{p}_i$. The new parameter values achieve the requested design change when the minimum is zero and partially achieve the change when it is positive. Our example design change can be achieved by increasing the circle radius, r_A, from 1 to 1.15 (figure 7.3d) or by decreasing the slot width, w_B, from 2.5 to 2.2 (figure 7.3e). Increasing r_A changes the entire contact space because the A circle is involved in every contact, whereas decreasing w_B only changes the channel width.

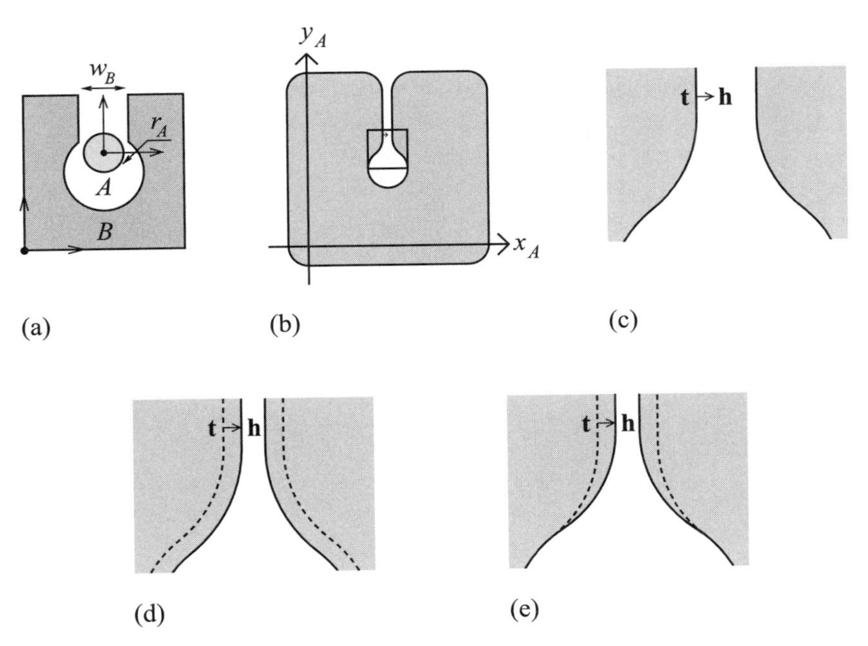

Figure 7.3
Circle-box pair: (a) initial design instance, (b) initial configuration space partition, (c) dragger, (d) $r_A = 1.15$ modification, (e) $w_B = 2.2$ modification.

7.2.2 Structure Change

Minimization of the dragger can change the free space structure. The structure is the number of free faces, the number of edge loops that bound each face, and the sequence of contact curves along each loop. A contact curve that enters or leaves a loop is a local change since the kinematic function is insensitive to which features generate it. A change in the free space topology implies a large change in kinematic function, which is often undesirable.

We illustrate these concepts with a driver-follower pair (figure 7.4). The initial partition of the configuration space shows a correct function (figure 7.4b). In the diagonal channel, the driver pin engages the follower slot and rotates the follower. In the horizontal channels, the driver arc aligns with the complementary follower arc and prevents rotation of the follower. The partition reveals excessive follower play in the driving mode. The play appears as the distance between the top and bottom contact curves that bound the channel (figure 7.4c). The designer requests smaller play with an upward dragger on the bottom curve. A local change in structure occurs

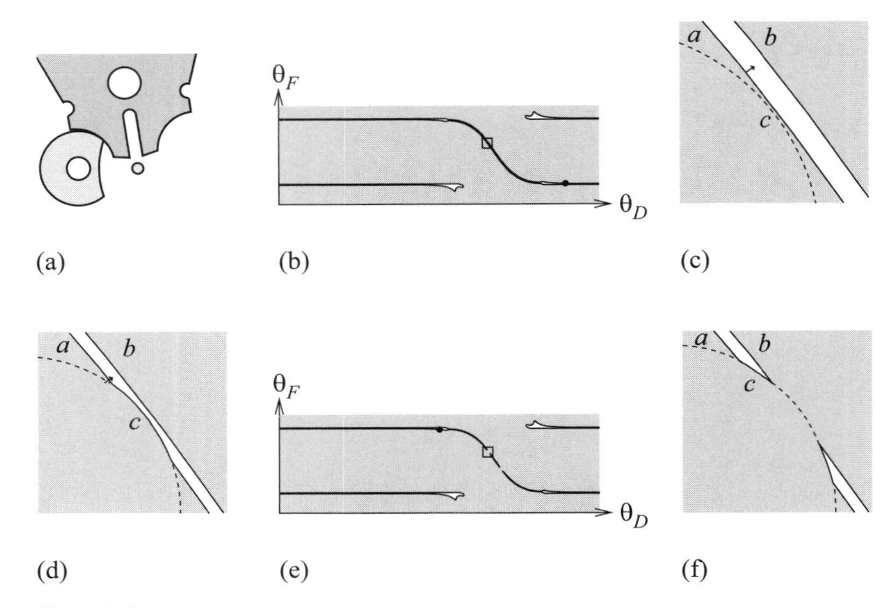

Figure 7.4
Driver-follower pair: (a) detail of initial instance, (b) partition, (c) detail with dragger, (d) local change, (e–f) topology change with detail.

partway through the optimization (figure 7.4d). The contact curve, *c*, of the driver arc-follower arc tip crosses the pin-slot curve, *a*, and enters contact space. A topology change occurs when *c* hits the upper channel boundary, *b*, and blocks the channel (figure 7.4e–f).

The parametric synthesis algorithm prevents topology changes. The new free space is matched against the old one. If the structures match, the new design instance is accepted. Otherwise the algorithm searches the line segment in parameter space between the old and the new instances for the first point, $\mathbf{p}_c$, where the structure changes. If the change is local, the optimization resumes from $\mathbf{p}_c$. In a topology change, two contact curves (on the same loop or on different loops) are disjoint on one side of $\mathbf{p}_c$, tangent at $\mathbf{p}_c$, and intersect on the other side of $\mathbf{p}_c$. The algorithm adds penalty terms to the objective function to prevent the tangency and restarts the optimizer at the old instance.

In our example, the tangent curves are *a* and *c* (figure 7.5a). One penalty term prevents *a* from crossing a barrier point below the tangency and the other prevents *c* from crossing a barrier above it. The barriers appear as a single filled circle because they are so close together (figure 7.5b). The

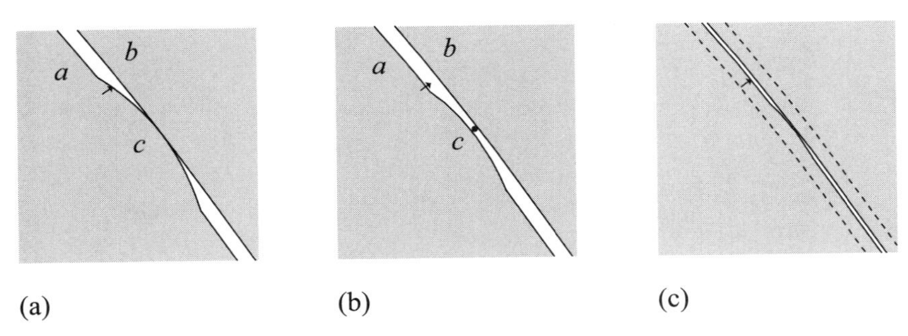

Figure 7.5
Topology change (a), barriers (b), and final partition (c).

revised objective is minimized and the design goal is achieved without changes in topology (figure 7.5c). The original channel (figure 7.4c) is drawn with dashed lines for comparison.

It remains to describe the matching algorithm. Two edges match if their contact curves are formed by the same pair of part features. For edge loops $e_1, \ldots, e_n$ and $f_1, \ldots, f_n$ to match, they must have the same length and e_1 must equal f_i for some i. The match succeeds when $e_2 = f_{i+1}, e_3 = f_{i+2}, \ldots,$ $e_n = f_{i+n}$ with the indices' modulo n. Partitions A and B match when they have the same number of free face edge loops and every A loop matches a single B loop. We can match the partitions by matching every A loop with every B loop. This naive algorithm, which takes $O(m^2)$ time with m the total number of edges, can be slow when m is large since it is used in the inner loop of the optimizer. A constant-time algorithm is cited in the notes.

7.2.3 Multiple Draggers

A design change can be specified with multiple draggers in multiple partitions. The objective function is the sum of one term, $|f_i(\mathbf{h}_i, \mathbf{p})|$, per dragger. The algorithm minimizes the objective as before while checking every pair for structure changes. We illustrate this using a mechanism with design flaws in two pairs, which are corrected with two draggers in one partition and one in the other.

The mechanism is the Dennis clutch from section 2.4 (figure 7.6). Initially, the pawl is disengaged from the ratchet and is locked by the arm (figure 7.6a). The operator releases the pawl by rotating the arm counterclockwise (figure 7.6b). The spring rotates the pawl clockwise until its tip engages a ratchet slot. The ratchet, which is mounted on a rotating shaft, drives the cam via the pawl (figure 7.6c). The rotating cam pin hits the upper finger of the arm (figure 7.6d) and rotates it clockwise. The lower

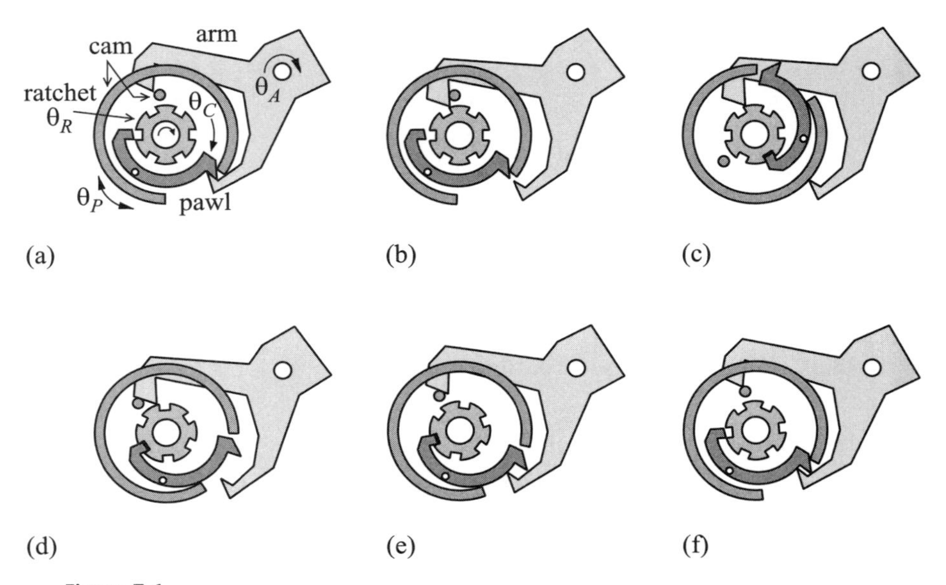

Figure 7.6
Dennis clutch mechanism: (a–f) configurations **a–f**.

finger of the arm hits the pawl (figure 7.6e), disengages it from the ratchet, and locks it (figure 7.6f). The cam rotates to its initial angle owing to inertia (figure 7.6a).

The designer achieves the intended function by parameter synthesis. The first change in design fixes a cam-arm design flaw: when the operator rotates the arm, it hits the cam arm before it can release the pawl. Partition of the cam-arm configuration space shows the design flaw (figure 7.7a). The initial configuration is **p** and the dashed line marks the angle at which the arm releases the pawl. The intended path for the motion is the vertical from **p** to the release line, but the path ends prematurely at **q** when the arm hits the cam's rim. The designer raises the contact curve with two draggers (figure 7.7b). The revised partition shows a correct function (figure 7.7c). The arm rotates from **p** to **q**, where it releases the pawl. The cam rotates from **q** to **r**, where its pin hits the upper finger of the arm. The cam rotates the arm from **r** to **s**, where the pin passes under the finger. The cam rotates to **p** to end the work cycle.

The second design change fixes a ratchet-pawl design flaw: when the pawl spring is too weak, the pawl tip can slide over a ratchet slot without engaging. A ratchet-pawl partition shows the design flaw (figure 7.8a). The intended motion path is from **p**, where the pawl tip reaches the edge of a

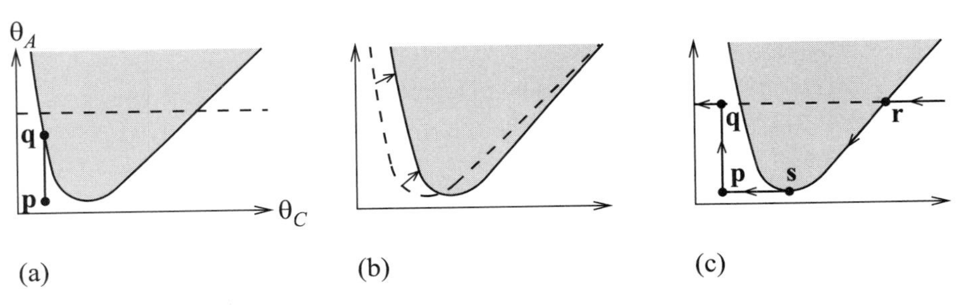

Figure 7.7
Cam-arm: (a) faulty partition, (b) synthesis, (c) correct partition.

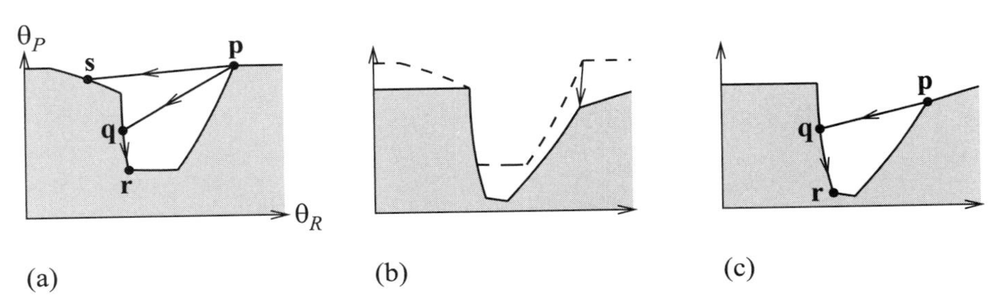

Figure 7.8
Ratchet-pawl: (a) faulty partition, (b) synthesis, (c) correct partition.

ratchet slot, to **q**, where it hits the slot side, to **r**, where it engages. The actual path is governed by the pawl's spring and the parts' inertias. If the spring is too weak, the pawl tip can pass the ratchet slot and hit the top of the next tooth at **s**. The designer lowers the right contact curve with a dragger (figure 7.8b). In the revised partition (figure 7.8c), the pawl tip hits the slot's side and engages independently of the spring's strength.

7.3 Parameter Optimization

The second synthesis algorithm is parameter optimization (figure 7.9). The input is a parametric model and a tolerance space. The goal is to ensure that every design instance has a correct kinematic function. An instance is deemed correct when its configuration space partitions match the nominal ones, as defined in section 7.2.2, and its motion path is within a specified distance from the nominal one. The algorithm searches the tolerance space for incorrect instances and then revises the nominal parameter values to exclude them.

Input: parametric model and tolerance space.
1. Compute partitions.
2. If no incorrect instances are found in tolerance space, exit.
3. Revise the nominal parameter values.
4. Go to step 3.
Output: revised nominal parameter values.

Figure 7.9
Algorithm for parameter optimization.

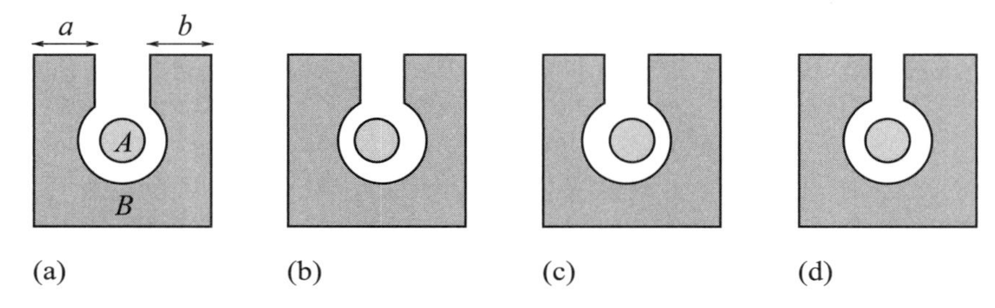

Figure 7.10
Circle-box pair: (a) nominal, maximal instances for (b) left, (c) right, and (d) both channel sides.

The algorithm does not optimize tolerances. We could tighten tolerances to exclude incorrect instances that cannot be excluded by changing the nominal parameter values. Tighter intervals are an inferior solution because they often increase manufacturing cost. We could also loosen tolerances to reduce cost. Optimization of tolerance requires a design quality metric, which is a separate research topic.

Maximal Instances We examine the instances that maximize the variation of one or more contacts. Variation in a contact represents a maximal variation in its kinematic function. Variation in two contacts represents a maximal interaction between them. Incorrect function is most likely to arise in such instances.

We illustrate maximal variation with a circle-box pair (figure 7.10). The parameters are $\mathbf{p} = (a, b)$ with nominal values $\overline{\mathbf{p}} = (2.75, 2.75)$. The circle's radius is 1 and the box length is 8. A maximal instance for the circle-left channel contact (figure 7.10b) is $(3.25, 2.75)$. A maximal instance for the circle-right channel contact (figure 7.10c) is $(2.75, 3.25)$. Both instances are correct in that the circle can pass through the channel (figure 7.10d). The instance $(3.25, 2.25)$ is maximal for both contacts and is incorrect.

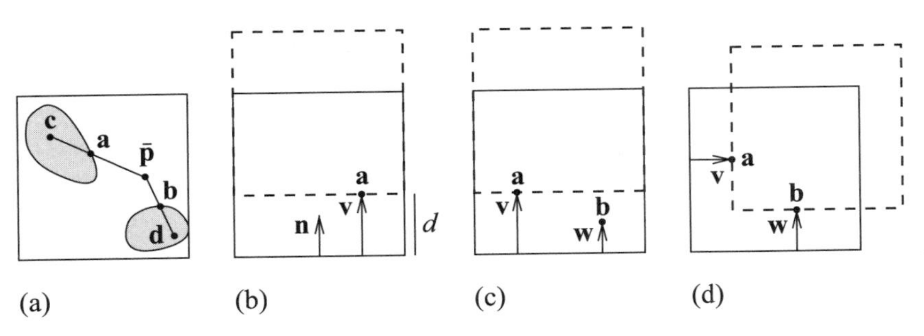

Figure 7.11
Parameter revision: (a) closest correct instances, (b) revision of one instance, and revision of two instances using one (c) and two (d) parameters.

The maximal instances are computed in two steps. The first step collects the parameter values that maximize the variation in the individual contacts. These are the values that generate the contact zone boundary points in step 3 of section 6.3. Each boundary point is determined by the values of the parameters in its contact equation. These values define a partial design instance. The other design parameters are called free. In our example, the partial instances are $(3.25, f)$ and $(f, 3.25)$ for the left and right channels, with f denoting a free parameter.

The second step forms the instances that maximize the variation in multiple contacts. Partial instances **a** and **b** are compatible when for every i either a_i is free, b_i is free, or $a_i = b_i$. They can be merged into a new partial instance whose ith element is free when a_i and b_i are free, equals a_i when b_i is free, and equals b_i otherwise. The merge assigns each partial instance its maximal parameter values; hence it maximizes the variation in both contacts. We merge all compatible sets of partial instances. We obtain the maximal instances by assigning the free parameters of the merges their nominal values. The two partial instances in our example are compatible and their merge, $(3.25, 3.25)$, is the joint maximum.

Revision The tolerance space is revised to exclude the incorrect maximal instances. Each incorrect instance is surrounded by a region of similar instances because its contact curves are continuous in the parameters. We would like to exclude the entire region, but computing it is impractical. Instead, we exclude the closest incorrect instance on the line segment from the nominal design to the incorrect instance, which we find by a bisection search. Figure 7.11a shows incorrect instances **c** and **d** with closest instances **a** and **b**; the incorrect regions are shaded.

We would like to compute a minimal revision of the nominal design that excludes the closest incorrect instances from the tolerance space. This task is computationally intractable, so we employ a heuristic. The minimal revision for a single instance, **a**, is computed as follows (figure 7.11b). Let d be the perpendicular distance from **a** to a tolerance space face with inward normal **n**. Incrementing the nominal parameter values by $\mathbf{v} = d\mathbf{n}$ excludes **a** from the tolerance space. The face that minimizes d yields the optimal revision (the dashed box).

We exclude multiple instances by combining their minimal revisions. Revisions **v** and **w** can be combined when v_i and w_i never have opposite signs. The ith element of the combination is $\max(v_i, w_i)$ when $v_i, w_i \geq 0$ and is $\min(v_i, w_i)$ otherwise. Figure 7.11c–d show combinations that revise one and two parameters.

7.4 Notes

Kinematic synthesis has been the subject of intense study for the past four decades. Most research has focused on linkages and other permanent contact mechanisms [25, 27, 57]. For a survey of earlier work, see Erdman [21]; a more recent survey is presented in McCarthy and Joskowicz [58]. Specialized techniques for synthesis of cam and gear mechanisms are described in Gonzales-Palacios and Angeles [26] and Litvin [51]. Software for synthesis of kinematic linkage includes Erdman's LINCAGES-2000, KINSYN, SyMech, and the ADAMS/View packages, among others.

Kinematic synthesis has also been studied in the broader context of mechanism design. Formal theories of mechanism design, including conceptual and configuration design, address these issues [29, 43]. Parameter synthesis as an optimization problem has gained much acceptance as an approach for optimal design [61] and has been applied to optimization of cam mechanisms [4].

The configuration space paradigm has been used for the synthesis of higher-pair mechanisms [12, 34, 49, 64]. Our work addresses parameter synthesis [45] and robust synthesis [46]. The constant-time partition-matching algorithm appears in Kyung and Sacks [45]. Robust synthesis couples nominal parameter and tolerance synthesis so that nominal and tolerance changes are evaluated together [76]. The nominal design is modified to reduce its sensitivity to variations of parts. Tolerances are then allocated to guarantee correct function and minimize cost.

The algorithms in this chapter extend in principle to general planar pairs, but they have not been implemented because of the complexity of manipulating three-dimensional configuration space.

8 Case Studies

This chapter presents four case studies of actual mechanisms taken from industry. They illustrate the use of the configuration space method in the analysis and design of mechanisms. Section 8.1 describes the kinematic synthesis of an optical filter mechanism. Section 8.2. describes the tolerance analysis of a selector mechanism for an automotive gearshift. Section 8.3 describes the redesign of a torsional ratcheting MEMS device. Section 8.4 describes the redesign of a spatial asynchronous reverse gear pair from an automotive transmission.

8.1 Optical Filter Mechanism

The optical filter mechanism [76] is from Israel Aircraft Industries and was introduced in section 7.1. The mechanism consists of a lens, a driver, and three filters mounted on identical followers (figure 8.1). The lens is attached to a fixed frame (not shown). The followers are stacked on a shaft and rotate independently. The driver consists of three slices that rotate together on a common shaft. Each driver slice drives the corresponding follower. The figure shows details of the top and middle driver slices and followers. Each slice consists of a driving pin and a locking arc. The second and third slices are rotated by 90° and 180° relative to the first slice. In the initial state, the filters are off the lens. When the driver shaft is rotated counterclockwise, the three followers are engaged in sequence. Each driver pin engages its follower slot and rotates the follower until the filter covers the lens. Rotating the driver clockwise resets the filters to the initial state.

The initial design scenario is discussed in section 7.1. The design task is to devise a mechanism to engage and reset the followers in the intended manner. The mechanism must be robust and compact because it will be mounted on a vehicle. The design concept is a Geneva mechanism with one driver and one follower per filter. The parametric model has 25

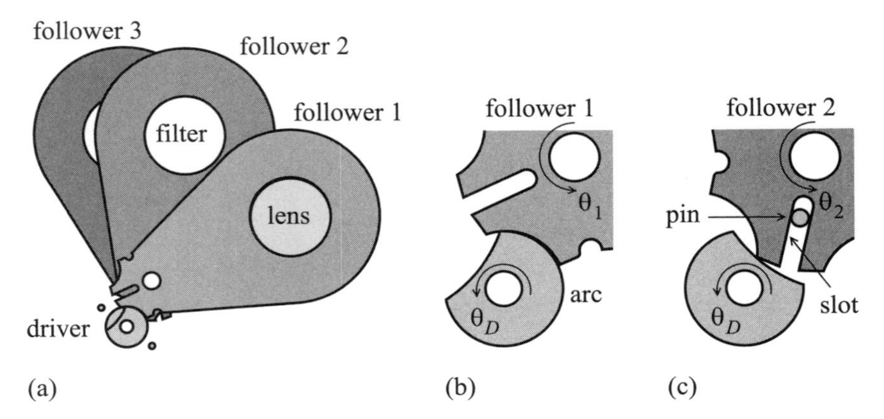

Figure 8.1
(a) Optical filter mechanism and (b–c) details of two driver-follower pairs.

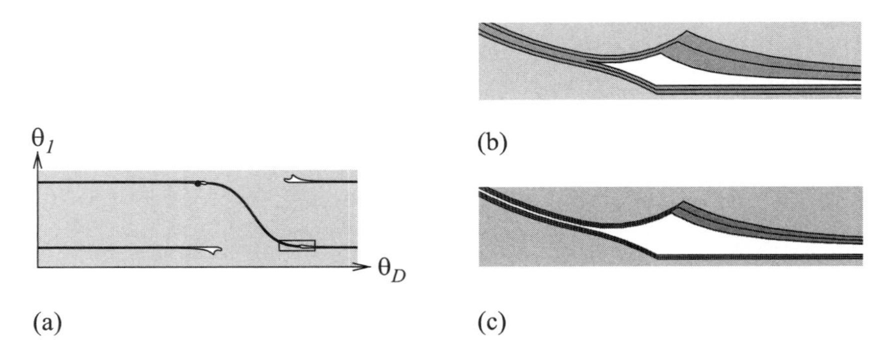

Figure 8.2
(a) Driver-follower configuration space partition with details of (b) original and (c) tightened contact zones.

functional parameters, including the centers of rotation, the pin and locking arc radii, and the slot dimensions.

We assign nominal parameter values that produce correct function using the synthesis algorithm from section 7.2. We assign tolerances to the parameters and assess their effects. Figure 8.2 shows a detail of the driver-follower contact zone in the area where the driver unlocks the follower and the pin enters the follower slot. The width of the contact zone varies with the sensitivity of the nominal contact configuration to the tolerance parameters. The upper and lower zones of the diagonal channel intersect, which suggests that the tolerance space contains design instances that block.

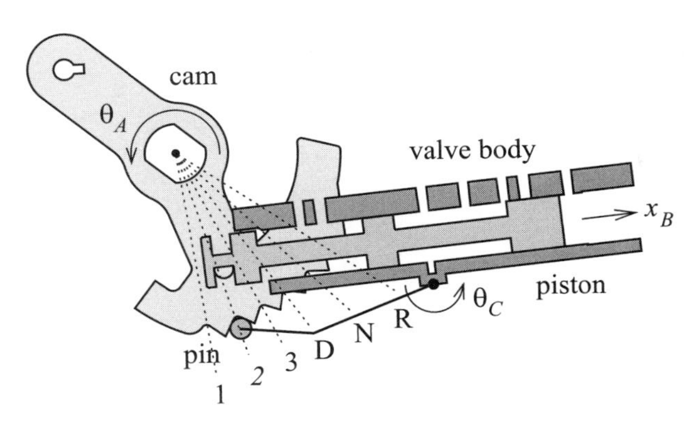

Figure 8.3
Gearshift mechanism.

We remove the blocking with the parameter optimization algorithm from section 7.3. The first iteration finds 146 blocking instances out of 700 candidates. After 19 iterations, the algorithm finds parameter values with no failures. If parameter optimization were to fail, we could tighten the tolerance intervals until the zones become disjoint, as shown in figure 8.2c.

8.2 Manual Transmission Gearshift

The gearshift selector mechanism [74] is from an an automotive automatic transmission from Ford Werke, Germany. The design task is to ensure correct function with manufacturing variation. Figure 8.3 shows the four main parts: a rotating cam, a translating piston, a rotating spring-loaded pin, and a fixed valve body. The cam pin fits in a slot in the left end of the piston, so cam rotation causes piston translation. The cam has seven slots that represent the seven gear settings: 1, 2, 3, D, N, R, and P (not shown). The automobile driver selects a setting by rotating the cam with the gearshift (not shown) to place the spring-loaded pin in the appropriate slot. The pin then locks the cam, which locks the piston. For each gear setting, the piston body opens and closes the appropriate conducts in the valve body. Variations in the pin, piston, and cam shapes and configurations affect the piston's displacement and thus the valve opening.

We construct a tolerance model by adding variation parameters to the functional features of the parts. For the cam, we tolerance the segments that form the slots and the pin that engages the piston. For the piston, we tolerance the two vertical line segments that are in contact with the cam

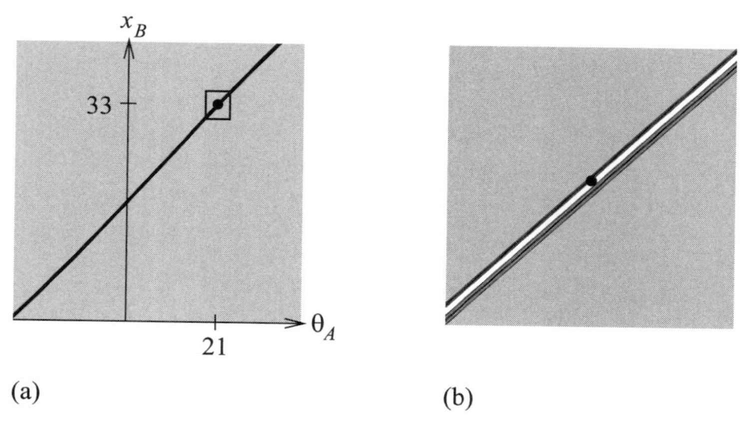

Figure 8.4
Cam-piston (a) configuration space partition and (b) detail of contact zone.

pin. For the pin, we tolerance the circle. Line segments are toleranced by varying the coordinates of the two endpoints; circle segments are toleranced by varying the radius and the center coordinates. To account for variability in the position of the rotation axes, we also tolerance the centers of rotation of the cam and the pin. Since we chose the piston as the reference part of the assembly, there was no need to tolerance the orientation of its translation axis. The model has 86 tolerance parameters for the cam, 8 for the piston, 5 for the pin, and 99 overall. We assign every parameter an independent tolerance of ± 0.1 mm.

Figures 8.4a and 8.5a show the configuration space partitions of the cam-piston and cam-pin pairs. The degrees of freedom are the cam orientation, θ_A, the piston offset along its axis of motion, x_B, and the pin orientation, θ_C. The cam-piston free space is a narrow diagonal channel whose top and bottom boundaries represent contacts between the cam pin and the left and right vertical line segments of the piston slot. The channel width quantifies the functional play. The kinematic relation is nearly linear. The cam-piston contact space consists of six "valleys" where the pin is in the cam slots, separated by "hills" where it switches slots.

We use tolerance analysis to determine the variation of the piston displacement for each cam setting. Figures 8.4b and 8.5b show details of the pair contact zones. The piston position, x_B, has a worst-case variation of 0.41 mm–0.45 mm and the pin orientation, θ_C, has a worst-case variation of 0.75°–1.03°.

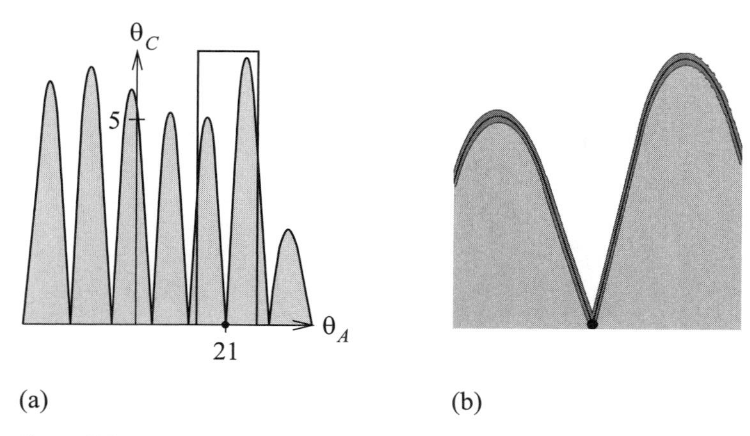

Figure 8.5
Cam-pin (a) configuration space partition and (b) detail of contact zone.

As described in section 6.4, we examine the sensitivity to the individual tolerance parameters at the nominal gear 3 setting $\theta_A = 21^\circ$, $\theta_C = 0^\circ$, $x_B = 33$ mm. The worst-case variation of x_B is 0.9 mm—roughly half from each pair. The main factors in the cam-piston variation are the cam center's horizontal position (25%) and vertical position (25%), tooth base x position (25%), and the x coordinates of the piston vertical segments (10%). The cam-pin variation is evenly distributed among the parameters of the touching features and the parts' centers of rotation.

We use parameter optimization to make the piston close the correct valves despite variation in its locked positions. The first iteration finds 331 design instances that allow excessive variation out of 715 candidates. After 150 iterations, the algorithm finds parameter values with no failures.

8.3 Torsional Ratcheting Actuator

The torsional ratcheting actuator [70] is a MEMS device designed and fabricated at Sandia National Laboratories Albuquerque, New Mexico. The design task is to remove intermittent blocking failures. The mechanism consists of a driver, a ratchet, a ring gear, and an antireverse (figure 8.6a). The gear and the antireverse are mounted on the substrate with pin joints and the ratchet is attached to the driver with a pin joint. The driver is attached to the substrate by springs that allow planar rotation but prevent translation. The driver is rotated 2.5° counterclockwise by an electrostatic

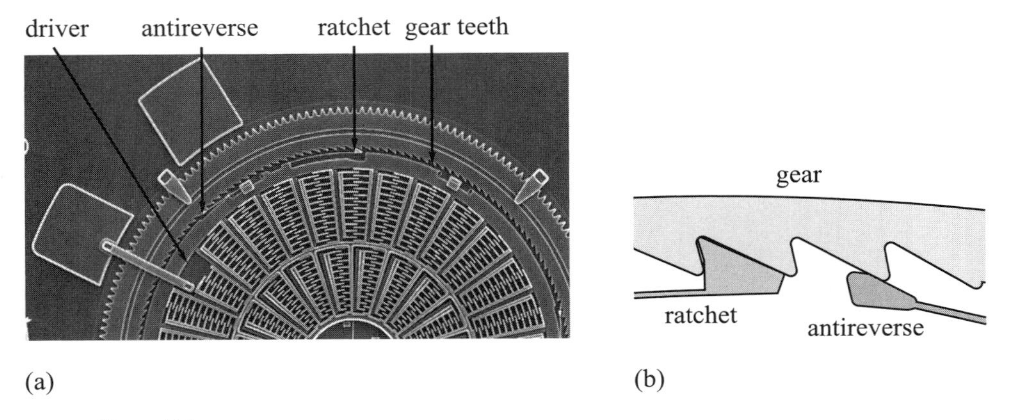

Figure 8.6
Torsional ratcheting actuator: (a) image courtesy of Sandia National Laboratories, (b) detail of the CAD model.

comb drive. The ratchet engages the inner teeth of the gear and rotates it counterclockwise. When the voltage drops, the springs restore the driver to its start orientation, which disengages the ratchet. The antireverse prevents the gear from rotating clockwise. Its outer teeth drive an external load.

We analyze the mechanism using the configuration space method (figure 8.7). The gear-antireverse partition shows that the pair functions correctly. The slanted contact curve that contains the displayed configuration is due to the long side of a gear tooth and the top of the antireverse. This contact causes the antireverse to follow the gear profile when the gear is driven. The steep contact curve to the left of this curve is due to the short side of a gear tooth and the side of the antireverse. This contact prevents the gear from rotating clockwise since every direction in which θ_G decreases lies in blocked space. The contact zone shows that both contact curves have the correct slope for every design instance.

The gear-ratchet pair has a three-dimensional configuration space because the ratchet rotates around a point on the driver and the driver rotates around a point on the frame. The contacts are invariant when the driver and the ratchet are rotated by the same angle, so we analyze the gear-ratchet pair in a two-dimensional space whose coordinates are the gear orientation and the angle between the driver and the ratchet. Equivalently, we fix the driver orientation at 0°, which fixes the ratchet's center of rotation. Figure 8.7e shows the configuration space partition. The near-vertical

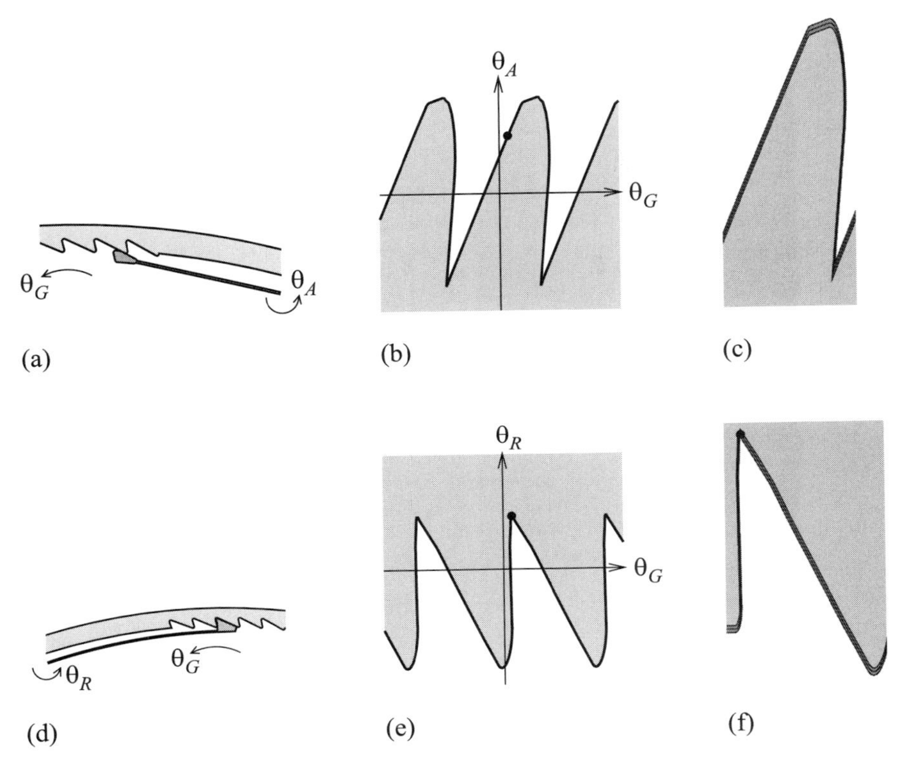

Figure 8.7
Initial design: (a–c) details of gear-antireverse pair, partition, and contact zone, (d–f) details of gear-ratchet pair, partition, and contact zone.

contact curve to the left of the displayed configuration is due to the short side of a gear tooth and the ratchet tip. This contact causes the gear to rotate with the driver. The contact curve to the right is due to the long side of the gear tooth and the ratchet back. This contact allows the ratchet to rotate clockwise and disengage the gear.

The contact zone (figure 8.7f) reveals a design flaw that causes intermittent failures: the near-vertical contact curve can have a positive slope in some design instances, which implies that the gear can rotate clockwise, escape the ratchet, and jump to the next tooth. The failure is intermittent because friction and inertia normally prevent jumps. We fix the design flaw with the parameter synthesis algorithm. In the revised design (figure 8.8), the gear tooth and ratchet slopes are larger, which makes the slope of the contact curve negative in every design instance.

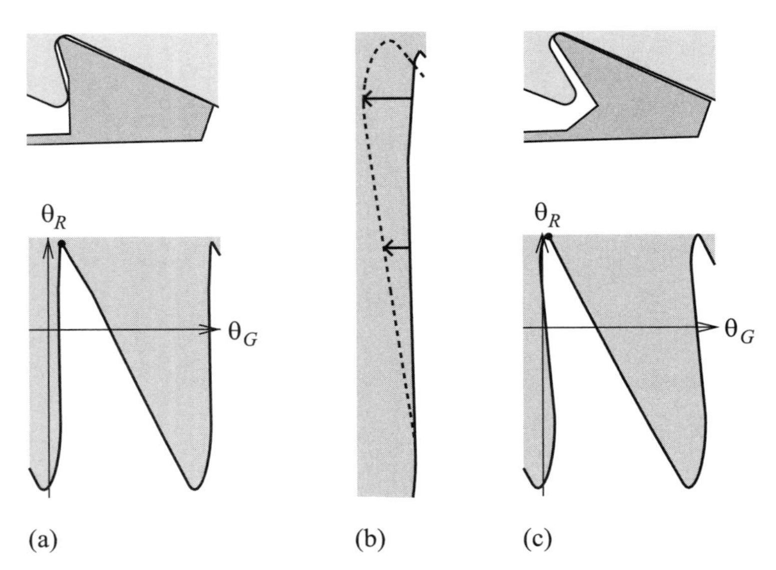

Figure 8.8
Gear-ratchet redesign: (a) before tooth and configuration space partition (detail), (b) draggers, (c) after.

8.4 Spatial Asynchronous Gear Pair

The spatial asynchronous reverse gear pair [75] is from an automotive transmission from Getrag-Ford Transmissions, Germany. The design task is to remove occasional gear blocking. The pair consists of two modified spur gears, the idler and the reverse gear, mounted on parallel axes (figure 8.9). The reverse gear rotates around its axis and the idler rotates around and translates along its axis. The idler has 31 teeth and the reverse gear has 13 teeth. Their pitch diameters are 65.62 mm and 27.52 mm. Figure 8.9b shows the geometry of the reverse gear tooth; the idler teeth have the same geometry with different dimensions. The tooth has involute sides, A and B, topped by spherical patches, C and D, that form a guiding chamfer. The spherical patches meet along a circular arc, e, and meet the involutes A and B along curves f and g. The tooth top is not functional.

The initial prototype usually functions correctly. When the automobile driver shifts into reverse, the idler translates toward the reverse gear until a pair of chamfers make contact (the configuration shown in figure 8.9a). The idler translation and the chamfer contact cause the the gears to rotate into alignment and to mesh. Figure 8.9b shows the contact point's path on the reverse gear tooth. The chamfer arcs e touch (point 1), then slide along

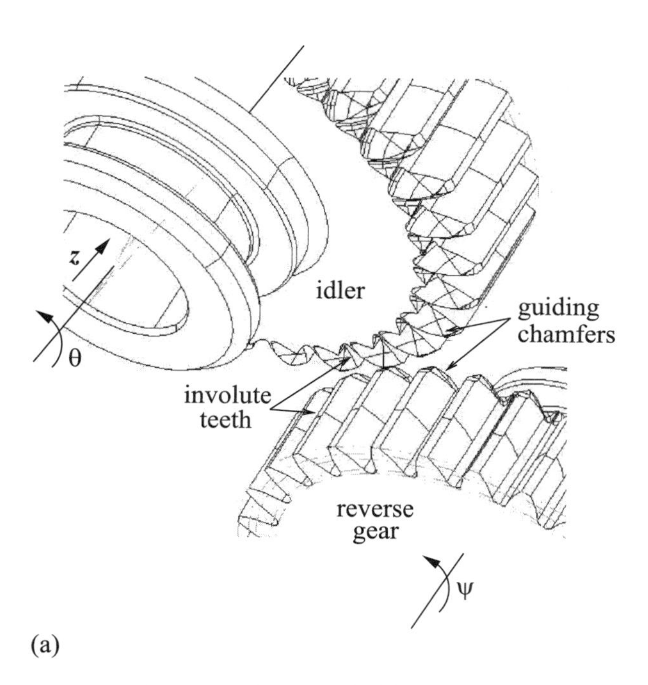

(a)

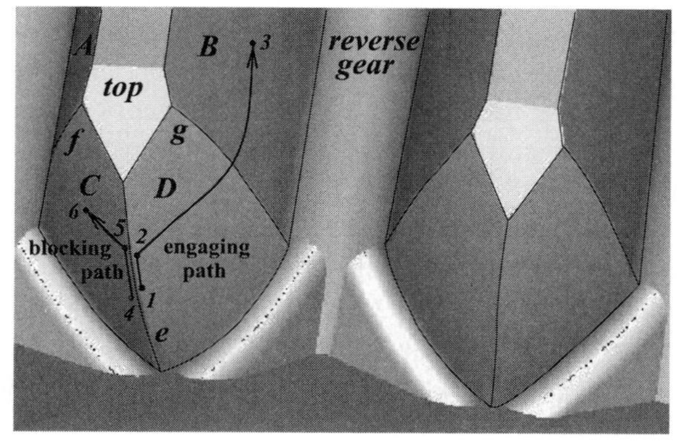

(b)

Figure 8.9
(a) Asynchronous gear pair and (b) geometry of reverse gear tooth.

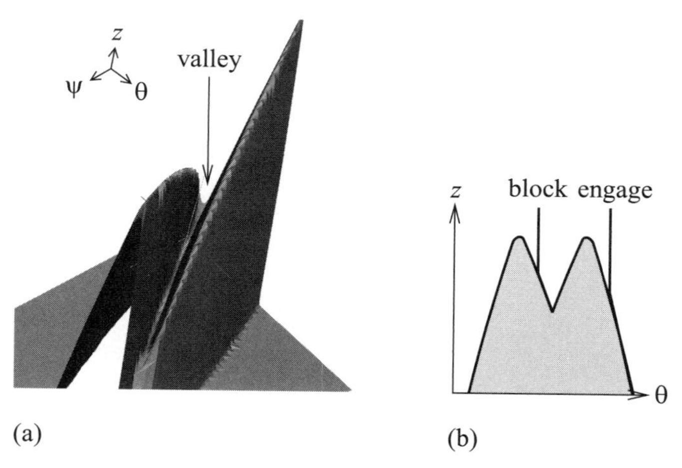

Figure 8.10
(a) Faculty configuration space partition, and (b) cross-section at $\psi = 0°$.

each other until the D patches touch (point 2). The D patch contact ends when the involutes B mesh (point 3).

The designers observed occasional blocking in which the idler stops translating before the gears mesh, owing to an interfering contact. Figure 8.9b also shows the path of the blocking contact point. The chamfer arcs e touch (point 4) then slide along each other until the C patches touch (point 5). The motion ends prematurely, owing to a contact between the C patches of an adjacent pair of teeth (point 6).

We use the configuration space method to find the initial configurations that lead to blocking and modify the design to eliminate blocking. The degrees of freedom are the idler rotation θ and translation z, and the reverse gear rotation ψ. Hence, the free space is three-dimensional and the contact space is two-dimensional. The contact space consists of two "hills" separated by a sloping "valley" (figure 8.10). We have not discussed the construction of three-dimensional partitions for spatial parts. The algorithm is analogous to the one for fixed-axis pairs, with contact curves and planar partitions replaced by contact surfaces and spatial partitions.

Figure 8.10b shows an engaging path and a blocking path in the $\psi = 0°$ cross-section. The driving motion is idler translation in the negative z direction. Initially, the idler is away from the reverse gear and the gear orientations θ and ψ can take any value. The idler translates (z decreases) until contact and then the gears rotate together. The idler engages if the path reaches $z = 0$ mm and blocks if the path ends in the valley. A path blocks

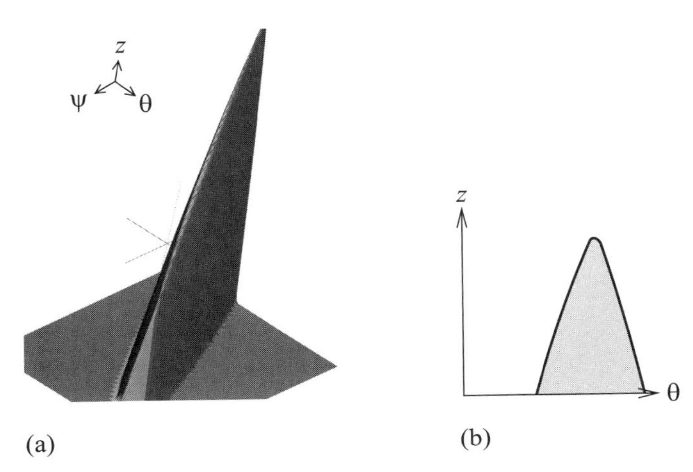

Figure 8.11
(a) Correct partition of configuration space (b) with cross-section.

based on its initial θ and ψ values. Measuring the contact space valley shows that 4% of paths block.

We eliminate blocking in two steps. A manual search of the design space reveals a design instance with 0.5% blocking. The contact space peaks move closer together, but the valley remains. The parameter synthesis algorithm might find a better design and would certainly be faster, but it has not been implemented for spatial parts. We eliminate blocking by removing every second idler guiding chamfer, which is a conceptual redesign. Figure 8.11 shows the partition of the modified design with the original dimensions of the parts. One peak has disappeared, so there is no valley and thus no blocking.

9 Conclusion

We have presented the configuration space paradigm for the kinematic design of mechanisms. This paradigm supports robust and optimal kinematic design of mechanisms by automating key tasks in the design cycle (figure 1.4). The main contribution of our work is support for higher-pair design. Our research strategy is to identify important mechanism classes, to study them in depth, and to develop efficient algorithms that support their design and analysis.

Configuration space is a complete, compact geometric representation of the motion and interaction of parts. It provides a principled approach to modeling the kinematics of mechanisms uniformly and systematically, including planar and spatial mechanisms with lower and higher pairs, contact changes, and topology changes. It covers nominal kinematics and kinematic variation that is due to manufacturing variation.

Within the configuration space paradigm, we have developed an efficient computational approach to the kinematics of contacts between parts. We systematically derived explicit, closed-form algebraic contact models for the most common planar feature contacts and outlined a general numerical contact model. We developed computational geometry algorithms for the construction of two- and three-dimensional configuration space partitions to model the contacts in kinematic pairs. The algorithms construct kinematic models from parametric descriptions of the parts' shapes and degrees of freedom.

Based on the representation of configuration space, we developed efficient algorithms to supports the core design tasks of analysis, tolerancing, and synthesis. For analysis, we developed algorithms for kinematic and dynamical simulation of mechanisms with changes in contact. For tolerancing, we developed algorithms for parametric worst-case tolerance analysis that are more accurate, efficient, and comprehensive than Monte Carlo simulation. For synthesis, we developed an interactive parameter synthesis algorithm and a parameter optimization algorithm. The algorithms are most useful for mechanisms with higher pairs and changes in the contact of parts. We presented case studies of how they have helped designers detect and correct unexpected behaviors and design flaws in automotive, MEMS, and optical mechanisms.

Research Directions

We conclude with development and research directions. The most important development goal is to integrate the configuration space method into a commercial CAD package (figure 1.1). This will allow designers to create and analyze mechanism models seamlessly within the configuration space paradigm. The main technical issues are software integration, converting CAD shape models to our format, and development of a graphical user interface. As a small first step, the prototype source code in appendix B provides a textual input format and a simple user interface. Another development direction is to integrate our kinematics software with multiphysics analysis, which will extend the configuration space method to nonrigid parts.

The main research directions are to increase the coverage of mechanism types and to improve the support for design tasks. The coverage of mechanisms can be extended by adding planar and spatial shape features of parts by developing contact models for the new features and by handling pairs with more degrees of freedom. For planar parts, the extension is to curve segments, such as splines and involutes. For spatial parts, the extension is to surface patches, such as helicoids, torii, and Bezier patches. Closed-form solutions for the contact equations will most likely rarely be available, so numerical solutions will be required.

Construction of a configuration space partition is implemented for fixed-axis pairs [42, 71]. Fast, robust construction of three-dimensional partitions is under development. General spatial pairs have six-dimensional partitions, which are impractical to construct. Reduction of dimensions, for example by slicing or projection, is a possible alternative. Construction of a mechanism partition is impractical for all but the simplest multipair mechanisms. The mechanism partition can be sampled by kinematic simulation. Other forms of local analysis should be considered, as should reduction of dimensions.

For tolerancing, the extensions include three-dimensional contact zones for general planar pairs and local contact zones for mechanisms. Statistical tolerancing can also be incorporated by considering the joint distribution of the tolerance parameters. The task is to compute the distribution of a kinematic function from the joint distribution of the tolerance parameters. In pair analysis, we compute a distribution for each contact zone region, whereas in mechanism analysis we compute a distribution along a nominal motion path. The kinematic function is linearized around the nominal parameter values. The linearization specifies the kinematic variation as a linear combination of given distributions. The outputs are the distributions of the kinematic variation in the contact zones and along the motion's path. Statistical tolerance analysis lays the groundwork for allocating tolerances based on cost functions. Finally, we have found that tolerance envelopes of the shapes of parts [60] are useful in quantifying functional errors, identifying unexpected collisions of parts, and determining if a mechanism can be assembled.

Kinematic synthesis is an open and fertile research area. Interactive modification using draggers can be extended to three-dimensional configuration spaces. Other interaction paradigms can also be contemplated. Support for configuration design and for incomplete, abstract specifications would also be valuable.

The systematic exploration and characterization of kinematic function is an important research topic. The goal is to develop a comprehensive taxonomy of this function. A promising avenue is to match kinematic function to configuration space partition regions and develop a symbolic and algebraic language to describe kinematic function [48]. The kinematic descriptions should characterize the qualitative and quantitative working of mechanisms concisely and abstractly. One approach is to summarize the function in terms of operating modes and mode transitions. Simplification and abstraction of kinematic function [37] is required to compare and classify mechanisms [38]. Based on this classification, a taxonomy of kinematic function could be developed and would provide a functional index for a database of mechanisms.

Visualization and interactive manipulation of configuration space partitions is another important area of research. The main challenges are to detect qualitative patterns in three-dimensional pair partitions and to visualize mechanism partitions. We can visualize the partitions of the kinematic pairs in a mechanism, which are projections of the mechanism partition, but it is difficult to determine the global topology from the projections. Intelligent projection techniques are needed to resolve this issue. The natural approach is to find subspaces that reveal the relevant interactions of parts and to visualize them in a perspicuous way.

Specific design tasks pose their own visualization problems. Simulation raises the need to visualize the motions of parts that generate paths in configuration space. One option is to animate the parts while tracing the corresponding configurations in the (full or projected) configuration spaces. This technique also helps probe the topology of the configuration space partition without constructing it. Tolerance analysis raises the need to visualize parametric families of partitions that represent the functional effects of manufacturing variations. Contact patches generalize to narrow zones surrounding the nominal patch surfaces.

The configuration space paradigm presented in this book may facilitate other contact analysis tasks that arise in mechanical engineering, robotics, computer graphics, and biomedical engineering. In mechanical engineering, contact analysis occurs in assembly [86] and [82] fixture design [10, 83] and part feeder design. In robotics, contact analysis occurs in path planning [69], analysis and planning of grasp and compliant motion, and robotic manipulation [54]. In computer graphics, contact analysis occurs in physics-based modeling and simulation, and in control and interaction in virtual environments. In biomedicine, contact analysis supports kinematic analysis of musculoskeletal action and joint and implant modeling [35].

Appendix A: Catalog of Mechanisms

This chapter contains a catalog of representative mechanisms and their configuration space partitions. It illustrates the relation between a part's shape, its configuration space partition, and kinematic function. The catalog consists of 24 higher pairs and 6 mechanisms containing 16 higher pairs. Of the 40 higher pairs, 33 are planar fixed-axis, 3 are general planar, and 4 are spatial fixed-axis. All the planar fixed-axis pairs and three of the mechanisms are available as HIPAIR files.

We begin each catalog entry with a description of the mechanism's structure and kinematic function. We then describe the configuration space partitions and explain their relation to the kinematic function. In some cases, we point out failure modes and establish the relation to other examples.

We use the following conventions for the figures. The left figure shows the parts' shapes, their typical configurations, configuration parameters, and motion arrows. The intended rotation of the part and translation directions are indicated with thick-headed arrows. The rotation angle is always measured in the standard counterclockwise direction, even when the arrow points clockwise. The translation is always measured along the standard right-hand coordinate axis, even when the arrow points left. A double thick-headed arrow indicates back-and-forth motion. The right figure shows the configuration space partition. For fixed-axis pairs, the horizontal and vertical axes are the driver and follower configurations. The part's configuration in the left figure is shown as a thick black dot. A kinematic simulation appears as a thick line. The simulation starts from the dot and lasts for one driver period. Unless otherwise specified, all angles are in $[-180°, 180°]$.

Contents

1 Offset Three-Arc Cam Pair

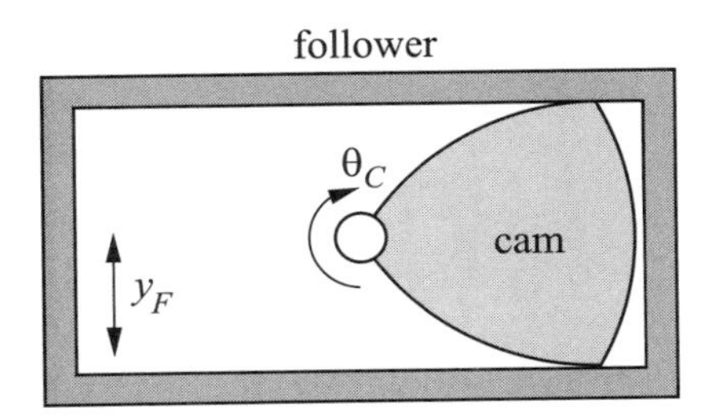

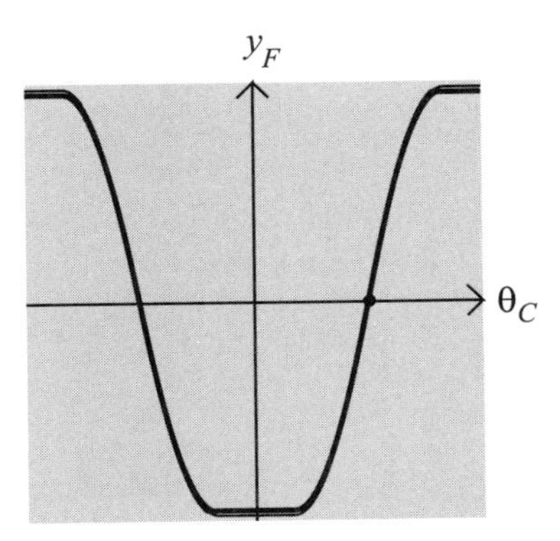

The pair consists of a rotating cam with orientation θ_C that drives a vertically translating follower with position y_F. The cam contour consists of three arc segments whose endpoints lie on a circle. The center of rotation coincides with the common endpoint of two incident arcs. The follower consists of a rectangular frame whose inner parallel horizontal line segments touch the cam.

A full rotation of the cam drives the follower up and down with two dwell periods in between. The dwell periods are at configurations where the common endpoint at the center of rotation touches the frame's upper and lower horizontal line segments. The contact point between the arc segment opposite the contact point and the follower lies on a circle of constant radius. The pair is bidirectional.

The configuration space partition consists of a single narrow channel whose nearly constant vertical height is the follower's backlash (play). The upper and lower horizontal channel segments correspond to the dwell period. The slanted left and right channel segments correspond to the driving periods. Its boundaries are sinusoidal curves.

2 Centered Three-Arc Cam Pair

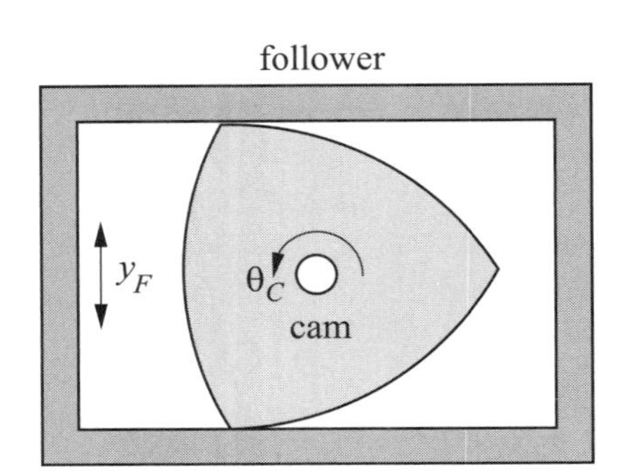

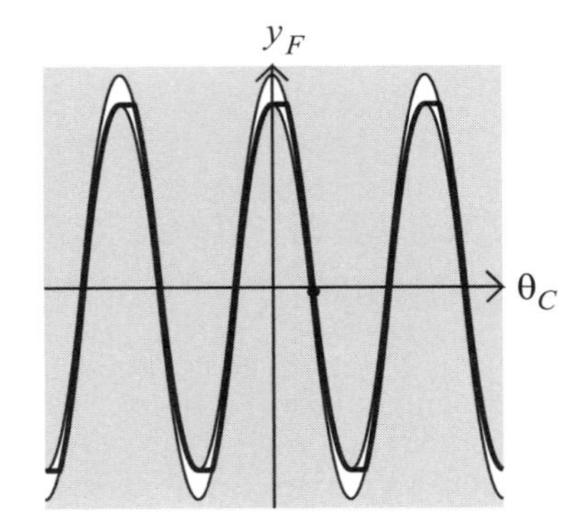

The pair consists of a rotating cam with orientation θ_C that drives a vertically translating follower with position y_F. The cam contour consists of three arc segments whose endpoints lie on a circle. Its center coincides with the cam's center of rotation. The follower consists of a rectangular frame whose inner horizontal line segments touch the cam. The distance between the upper and lower contact points is constant, so the cam fits tightly inside the follower without interference.

A full rotation of the cam drives the follower up and down three times at evenly spaced intervals. In each interval, one cam arc segment touches one horizontal follower line segment. The transitions occur at the cam's endpoints. The pair is bidirectional.

The configuration space partition consists of a single narrow channel whose vertical width is the follower's backlash. The slanted left and right channel segments correspond to the driving periods, whose boundaries are sinusoidal curves. The short horizontal segments represent the dwell periods when the cam rotates inside the follower without pushing it. The rising and falling segments represent the periods when the cam pushes the follower right and left. The cam arc segment breaks contact with one side of the follower at the start of each dwell period. The next arc segment makes contact with the opposite vertex of the cam at the end of the dwell period. The vertical channel width is the follower's backlash. Although the pair is structurally similar to pair 1 (the offset three-arc cam pair), the change in the cam rotation center yields a different kinematic function.

3 Six-Arc Cam Pair

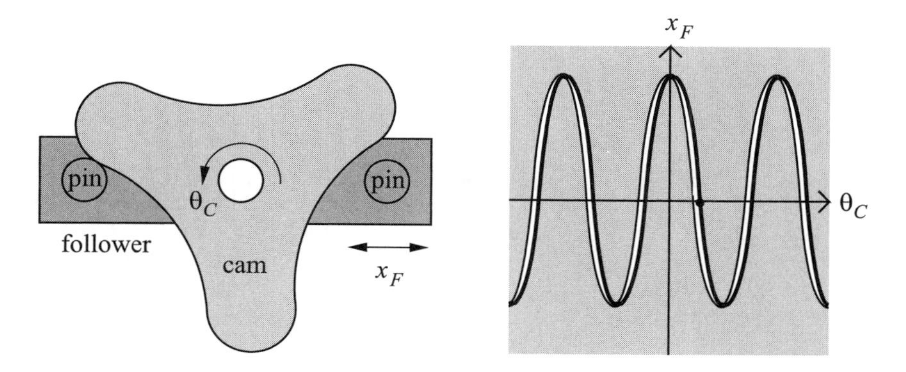

The pair consists of a rotating cam with orientation θ_C that drives a horizontally translating follower with position x_F. The cam's contour consists of three convex circular arc lobes connected by three concave tangent circular arcs. The follower consists of two circular pins mounted on a rectangular frame.

A full rotation of the cam drives the follower left and right three times at evenly spaced intervals. The cam lobes alternately push the left and right pins. Since the distance between the left and right contacts is constant, the cam fits tightly between the two pins without interference. The pair is bidirectional.

The configuration space partition consists of a single narrow channel whose nearly uniform vertical width is the follower's backlash. The slanted left and right channel segments correspond to the driving periods, whose boundaries are sinusoidal curves. The rising and falling segments of the kinematic simulation represent the periods when the cam pushes the follower right and left. Pair 2 has a similar function, but this pair has less backlash and no dwell.

4 Unidirectional Three-Finger Cam Pair

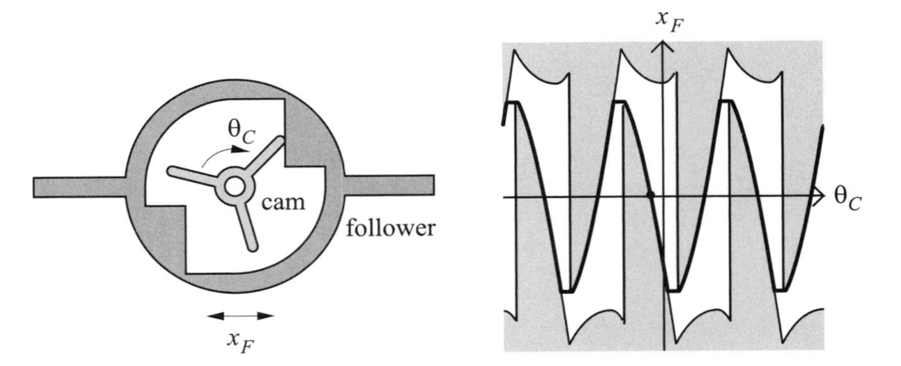

The pair consists of a rotating three-finger cam with orientation θ_C that drives a horizontally translating follower with position x_F. The cam finger tips are half-circles. The inner follower boundary is mirror symmetric. It consists of an outer arc circle, an upper horizontal segment, a vertical segment, and a middle horizontal segment.

A full clockwise rotation of the cam drives the follower left and right three times, with short dwell periods in between. The cam finger tips alternately push the upper vertical segment of the follower right and the lower vertical segment left. Counterclockwise rotation of the cam causes it to block when one of its finger line segments hits a horizontal segment of the follower. The pair is thus unidirectional.

The configuration space partition consists of a single channel with shifted, mirror-symmetric upper and lower boundaries. The slanted channel boundary segments correspond to driving periods. The vertical channel boundary segments correspond to blocking configurations in which the cam cannot rotate. The kinematic function follows the thick line from right to left. The horizontal segments represent the dwell periods when the cam does not touch the follower. The rising and falling segments are sinusoidal curves that represent the periods when the cam pushes the follower right and left. A cam finger breaks contact with one side of the follower at the start of each dwell period. The next finger makes contact with the other side of the cam at the end of the dwell period. Its clockwise kinematic function is similar to that of pairs 2 and 3.

5 Bidirectional Three-Finger Cam Pair

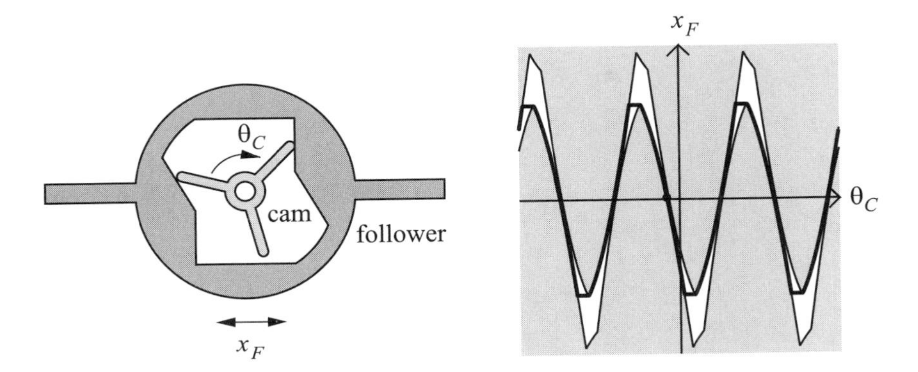

The pair consists of a rotating three-finger cam with orientation θ_C that drives a horizontally translating follower with position x_F. The cam finger tips are half-circles. The inner follower boundary is mirror symmetric. It consists of an outer arc circle, an upper horizontal segment, a vertical segment, and a middle slanted segment.

A full clockwise rotation of the cam drives the follower left and right three times with short dwell periods in between. The cam fingers alternately push the upper vertical segment of the follower right and the lower vertical segment left. A full counterclockwise rotation of the cam drives the follower in the same way, but the cam fingers now push the lower slanted segment right and the upper ones left. The pair is bidirectional.

The configuration space partition consists of a single narrow channel with shifted, mirror-symmetric upper and lower boundaries. The slanted channel segments correspond to driving periods. The horizontal segments of the kinematic function line represent the dwell periods when the cam does not touch the follower. The rising and falling boundary segments are sinusoidal curves that represent the periods when the cam pushes the follower right and left (or left and right). A cam finger breaks contact with one side of the follower at the start of each dwell period. The next finger makes contact with the other side of the cam at the end of the dwell period. The free space channels are narrower than those of pair 4. The configuration space partition is similar to those of pairs 2 and 3.

6 Geneva Pair

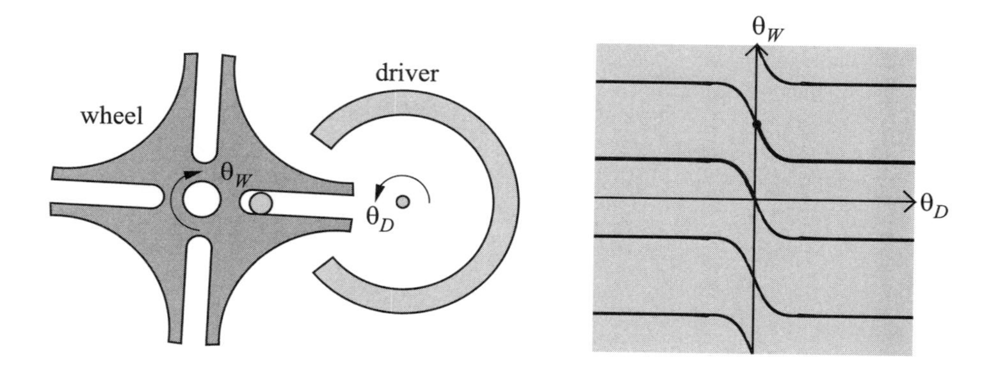

The pair consists of a rotating driver with orientation θ_D and a rotating Geneva wheel with orientation θ_W. The driver consists of a driving pin and a concentric locking arc segment mounted on a disk (not shown). The Geneva wheel consists of a cylindrical basis with four slots on which four locking arc segments are symmetrically mounted.

As the driver rotates, the pin engages the Geneva wheel slots and rotates it nonuniformly in the opposite direction. The locking arc segments of the driver engage those of the wheel during the dwell periods, thereby preventing the wheel from rotating. A full rotation of the driver causes a nonuniform, intermittent rotation of the wheel with four dwell periods and four drive periods. The pair is bidirectional.

The configuration space partition consists of a single narrow channel whose nearly uniform vertical width is the wheel's backlash. The slanted channel segments correspond to the driving periods, whose boundaries are sinusoidal curves.

7 Inverse Geneva Pair

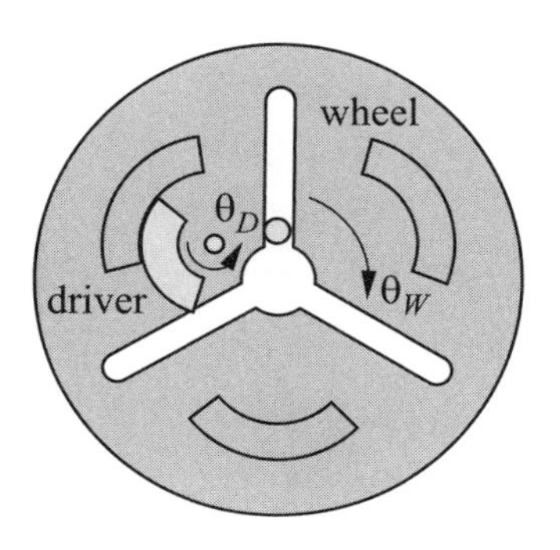

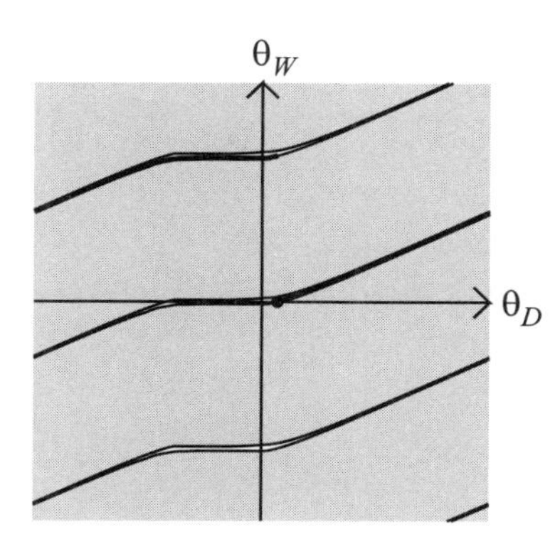

The pair consists of a rotating driver with orientation θ_D and a rotating inverse Geneva wheel with orientation θ_W. The driver consists of a driving pin and a concentric locking arc segment. The Geneva wheel consists of a cylindrical base with three slots on which three locking arc segments are symmetrically mounted.

As the driver rotates, the pin engages the inverse Geneva wheel slots and rotates the wheel in the same direction. The locking arc segments of the driver engage those of the wheel during the idle periods, thereby preventing the wheel from rotating. A full rotation of the driver causes a nearly uniform, intermittent rotation of the wheel, with three dwell periods and three drive periods. The pair is bidirectional.

The configuration space partition consists of a single narrow channel whose nearly uniform vertical width is the wheel's backlash. The slanted channel segments correspond to the driving periods, whose boundaries are sinusoidal curves. Compared with the pair 6 partition, the channels are slanted in the opposite direction and the driving periods are longer.

8 Intermittent Cam Pair

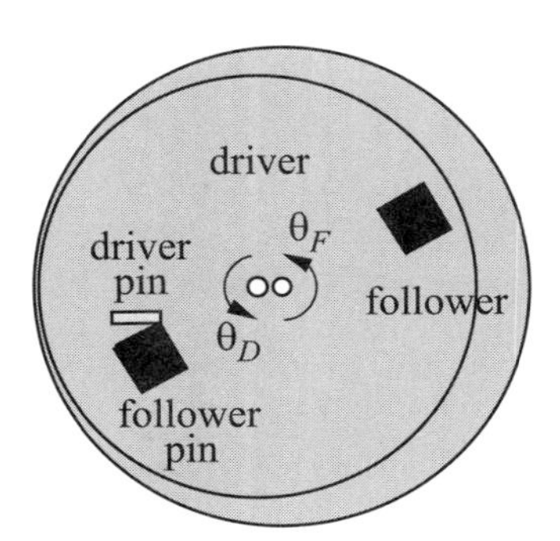

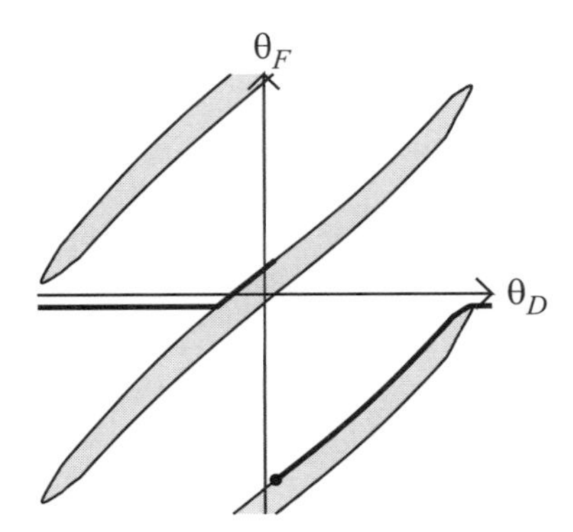

The pair consists of a rotating driver with orientation θ_D and a rotating follower with orientation θ_F. The driver consists of a rectangular pin mounted on a base. The follower consists of two square pins symmetrically mounted on a disk. The centers of rotation for both parts are offset.

As the cam rotates, its rectangular pin pushes the follower square pin for half a turn. The pins then slide out of engagement because of the eccentricity of the axes of rotation. The follower is idle during the second half of the cam rotation. The rectangular pin of the cam then engages the second square pin of the follower for another half turn. The pair is bidirectional.

The configuration space partition consists of two slanted elongated strips that nearly span a full turn. The strip boundaries correspond to the contact of the driver pin with one of the square follower pins. The kinematic simulation has a horizontal segment in free space where the driver rotates and the follower is idle, and a slanted segment where the driver rotates the follower for nearly half a turn.

9 Interlock Pair

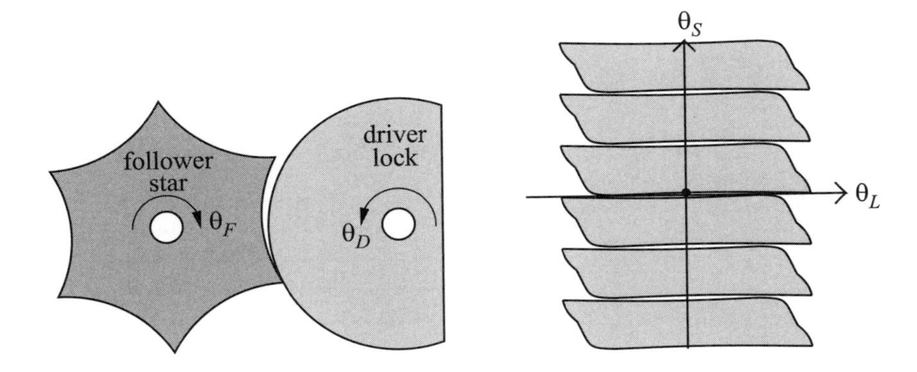

The pair consists of a rotating driver lock with orientation θ_D and a rotating star-shaped follower with orientation θ_F. The driver lock consists of an arc segment centered at the center of rotation and of a line segment. The follower star consists of six arc segments whose centers coincide with the center of rotation of the lock.

The driver lock locks the follower star at six orientations. The orientation of the follower star can be changed by first turning the lock so that the line segment faces it vertically, then rotating it. When the follower star is spring loaded one way and the driver rotates the other way, the follower turns one-sixth of a turn for each rotation of the driver lock. The pair is bidirectional.

The configuration space partition consists of six identical slabs that span 360°. The narrow horizontal channels between slabs correspond to configurations in which the driver rotates and the follower is locked. At the end of the channel, the follower star turns until it enters the next horizontal channel.

10 Star Indexer Pair

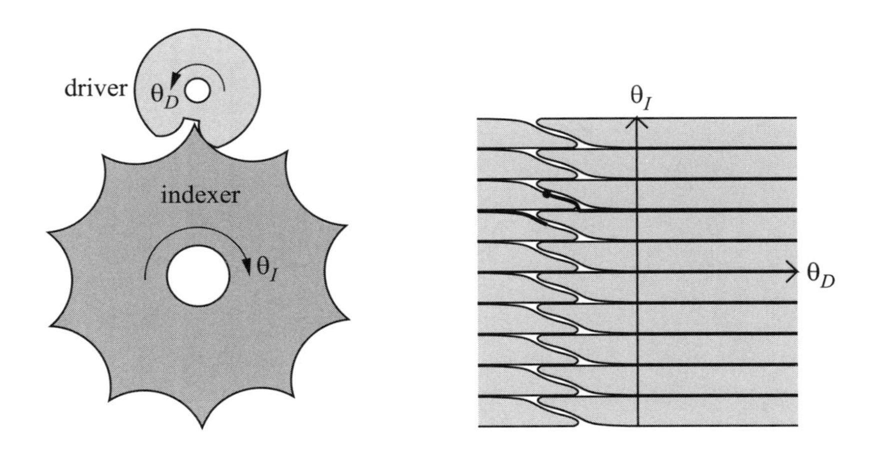

The pair consists of a rotating driver with orientation θ_D and a rotating star-like indexer with orientation θ_I. The driver consists of a disk with a recess that allows the cusps of the indexer arc segment to pass underneath. The indexer consists of ten arc segments whose centers coincide with the center of rotation of the driver. The indexer is spring loaded counterclockwise.

A full counterclockwise rotation of the driver advances the indexer clockwise by one-tenth of a turn and locks it for the rest of the rotation. If the indexer were not spring loaded counterclockwise, it would not engage the driver slot and thus would not advance. If the spring is too weak, the driver slot passes the indexer pin before it can engage, so the indexer does not advance. The pair is unidirectional.

The configuration space partition consists of ten horizontal and ten slanted narrow channels. The horizontal channels correspond to the dwell periods of the indexer. The slanted channels correspond to the driving periods. The channel width is the backlash between the driver and the indexer, which is largest at transitions between drive and dwell periods. The configuration space partition, and thus the kinematic function, is nearly identical to that of pair 11.

11 Disk Indexer Pair

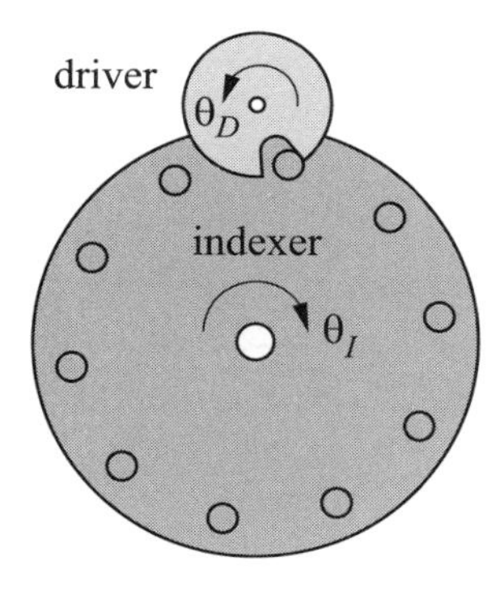

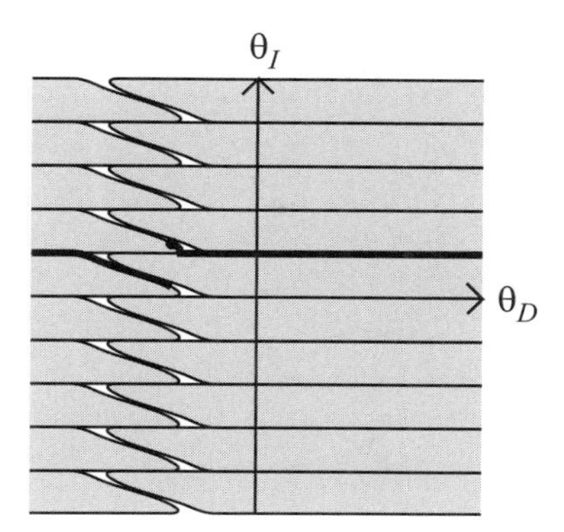

The pair consists of a rotating driver with orientation θ_D and a rotating indexer with orientation θ_I. The driver is a disk with a slot cutout that engages the indexer pins. The indexer has ten circular pins evenly spaced on a disk and is spring loaded clockwise.

Continuous counterclockwise rotation of the driver causes intermittent clockwise rotation of the indexer. In drive periods, the driver slot engages one indexer pin and advances the indexer one-tenth of a turn. In dwell periods, the outer arc of the driver engages two adjacent indexer pins and prevents rotation of the indexer. When a dwell period ends, the indexer spring rotates the next indexer pin into the driver slot and the next drive period begins. If the indexer were not spring loaded, its configuration would stay constant as the driver rotates. If the spring is too weak, the driver slot passes the indexer pin before it can engage, so the indexer does not advance.

The configuration space partition consists of ten horizontal and ten slanted narrow channels. The horizontal channels correspond to the dwell periods of the indexer. The slanted channels correspond to the driving periods. The channel width is the play between the driver and the indexer, which is largest at transitions between drive and dwell periods. The configuration space partition, and thus the kinematic function, is nearly identical to that of pair 10.

12 Lever Indexer Pair

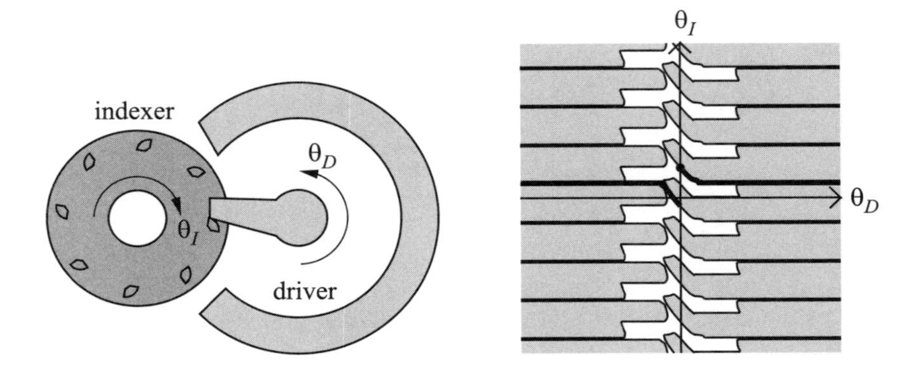

The pair consists of a rotating driver with orientation θ_D and a rotating indexer with orientation θ_I. The driver consists of a driving finger and a locking arc mounted on a disk (not shown). The indexer has eight trapezoidal pins evenly spaced on a circular base.

Continuous counterclockwise rotation of the driver causes intermittent clockwise rotation of the indexer. In drive periods, the driver pin engages one indexer pin and advances the indexer one-eighth of a turn. In dwell periods, the locking arc engages two adjacent indexer pins and prevents the rotation of the indexer. The pair is unidirectional.

The configuration space partition consists of eight horizontal narrow channels and eight free space regions in between. The horizontal channels correspond to the dwell periods of the indexer. The channels' width is the play between the driver and the indexer. The free space regions have a narrow channel. The kinematic simulation consists of a slanted segment where the driver pushes the indexer, followed by a long horizontal segment, followed by a short slanted segment. Unlike pairs 10 and 11, spring loading the indexer is unnecessary. The configuration space partitions are similar.

13 Interlock Drive Pair

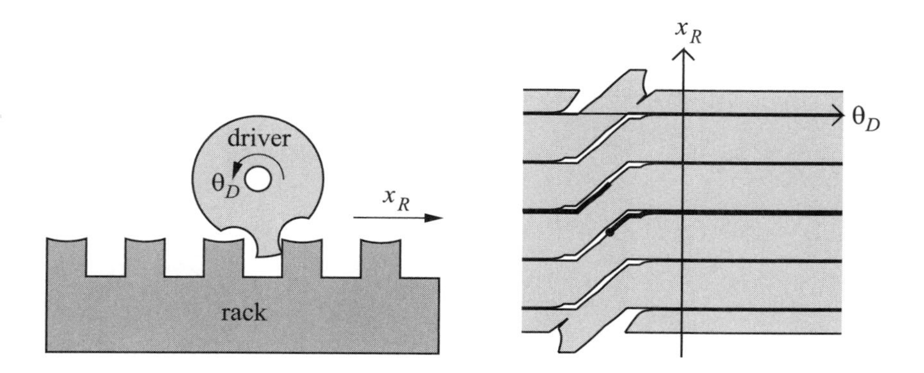

The pair consists of a rotating driver with orientation θ_D and a horizontally translating rack with position x_R. The driver is a disk with a single tooth and two cutouts to allow the follower teeth to pass. The rack has five teeth. The tooth top is a concave arc segment whose radius is the same as that of the driver disk.

At each driver rotation, its tooth engages a rack tooth and pushes it. Then its locking arc engages the concave top of the tooth and locks the rack. Continuously rotating the driver makes the rack translate with dwell periods. The pair is bidirectional.

The configuration space partition consists of four horizontal and four slanted channels. The horizontal channels correspond to the dwell periods of the indexer. The channels' width corresponds to the play between the driver and the indexer. The top and bottom channels represent contacts with the leftmost and rightmost tooth. Pair 7 has a similar partition.

14 Ratchet Pair

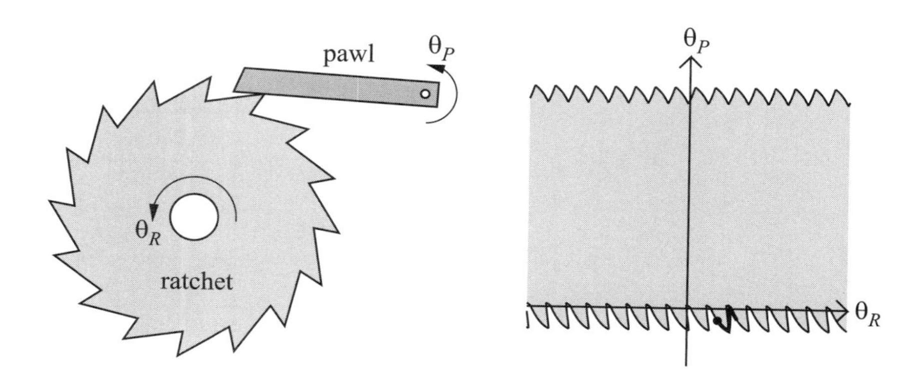

The pair consists of a rotating ratchet with orientation θ_R and a rotating pawl with orientation θ_P. The ratchet consists of 16 triangular teeth. The pawl is a quadrangle whose slanted tip matches the shape of the ratchet teeth.

Rotating the ratchet counterclockwise causes the pawl to rise and then fall under its own weight, following the profile of the teeth. The pawl blocks clockwise rotation of the ratchet after at most one-sixteenth of a rotation. Being unidirectional is the primary kinematic function.

The configuration space partition consists of an upper and lower boundary. The lower boundary corresponds to configurations where the pawl is above the ratchet, as shown in the figure. The upper boundary corresponds to configurations where the pawl side is in contact with the ratchet teeth to the right of the ratchet. The lower boundary realizes the kinematic function. It has one notch for each tooth of the ratchet. The notch slanting prevents simultaneous clockwise rotation of the ratchet and counterclockwise rotation of the pawl.

15 Escapement Pair

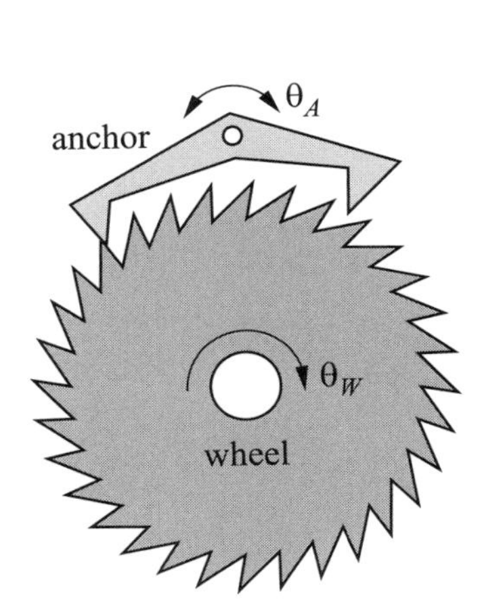

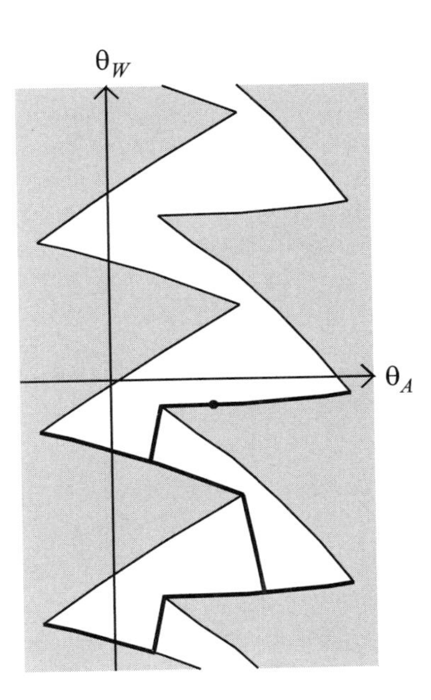

The pair consists of a rotating wheel with orientation θ_W and a rotating anchor with orientation θ_A. The wheel consists of 30 slanted triangular teeth. The anchor consists of left and right triangular pallets that alternately engage the wheel teeth. Attached to the anchor is a pendulum (not shown) that oscillates with a constant period. The wheel rotates counterclockwise, owing to a torque acting on its hub.

As the anchor oscillates, the pallets alternately engage teeth on the left and right sides of the wheel. The wheel advances one tooth each oscillation. It rotates at a constant velocity that is determined by the period of the anchor. The timing is such that the wheel advances one tooth every second.

The figure shows a detail of the configuration space partition. The upper and lower boundaries correspond to contacts of the right and left anchor pallets with the teeth. The interleaved boundaries prevent the wheel from rotating freely. In the kinematic simulation curve, the nearly horizontal segments on the boundary show that the anchor oscillates back and forth while the wheel follows it. The nearly vertical segments in free space show that the wheel rotates quickly in between.

16 Double-Rack Gears Pair

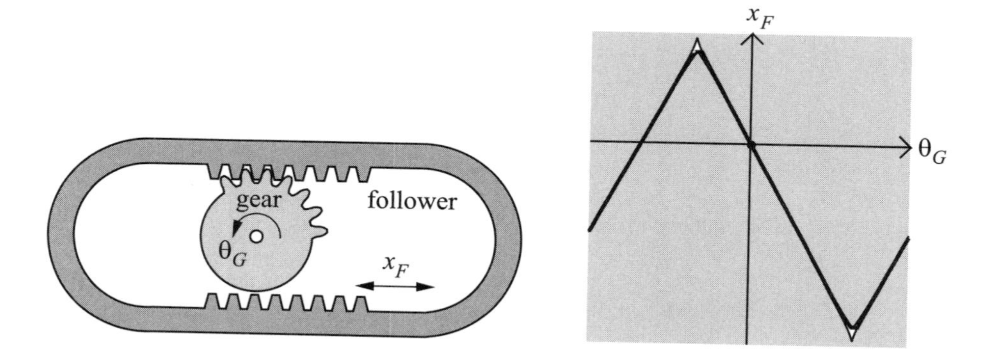

The pair consists of a rotating sector gear with orientation θ_G and a horizontally translating follower with position x_F. The sector gear has teeth that span less than half of its circumference. The follower has geared racks at the top and the bottom of its inner perimeter. The upper and lower racks are offset with respect to each other. The sector gear alternately engages the upper and lower racks, pushing the follower left and right.

Continuous rotation of the sector gear causes the follower to move left and right without dwell. The maximum displacement of the follower is equal to the perimeter of the sector gear. The pair is bidirectional.

The configuration space partition consists of a narrow channel with three nearly linear segments: two right slanted and one left slanted. The channel width is the play between the driver and the indexer. The slanted segments correspond to the right and left motions of the follower.

17 Sector Gears Pair

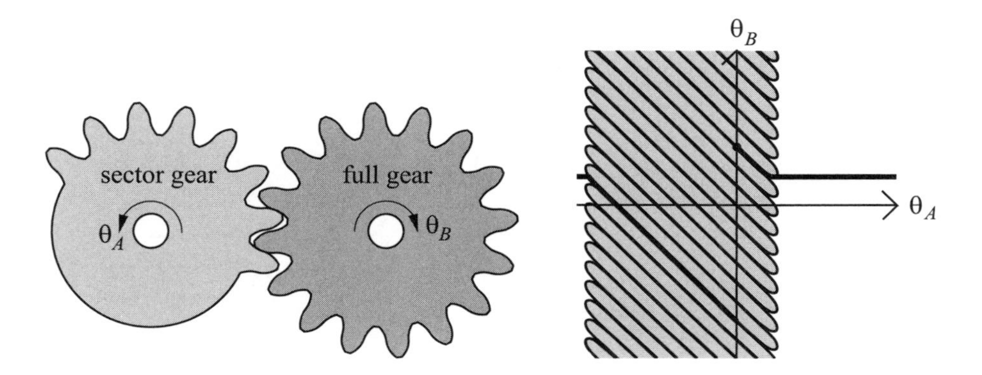

The pair consists of a rotating sector gear with orientation θ_A and a rotating full gear with orientation θ_B. The gears have equal radii. The sector gear has 8 teeth, while the full gear has 16. When the gears are engaged and one gear rotates in either direction, the other gear rotates in the opposite direction by the same angle. Two pairs of teeth are engaged at all times. The transmission ratio is almost linear, with slight backlash and chatter. When the gears are disengaged, their rotations are independent.

Continuous rotation of the sector gear causes the full gear to alternately rotate half a turn and dwell. Continuous rotation of the full gear causes the half gear to rotate in the opposite direction by at most half a turn. The pair is bidirectional.

The configuration space partition consists of 16 slanted narrow channels and a free space region (right). The slanted channels correspond to the driving periods at each of the 16 possible relative orientations in which the gears are engaged. The channel width is the play between the two gears. The horizontal segment of the kinematic simulation represents the dwell period when the sector gear turns and the full gear is at rest. The slanted line segment represents the gears rotating in opposite directions.

18 Dwell Gears Pair

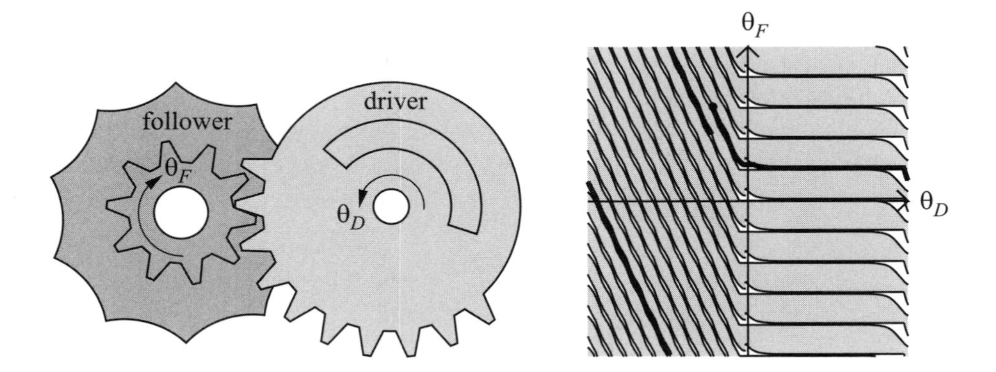

The pair consists of a rotating driver sector gear with orientation θ_D and a rotating follower gear with orientation θ_F. The driver is a concentric arc segment mounted on a gear sector with ten trapezoidal teeth. The follower is a star-shaped disk mounted on a gear wheel. It has ten outer locking arc segments and ten teeth. The pair is a combination of pairs 9 and 17.

As the driver rotates counterclockwise, the driver sector gear engages and turns the gear wheel clockwise. When the sector disengages, the concentric arc segment engages one of the ten concave arcs of the star-shaped disk, thereby locking it in a dwell position. A full rotation of the driver advances the follower by one-tenth of a rotation. The pair is bidirectional.

The configuration space partition consists of ten horizontal and ten slanted narrow channels. The horizontal channels correspond to the dwell periods of the indexer. The slanted channels correspond to the driving periods. The channel width is the play between the driver and the follower gears, which is largest at transitions between drive and dwell periods. The configuration space partition is similar to those of pairs 9–12.

19 Double-Sector Gears Pair

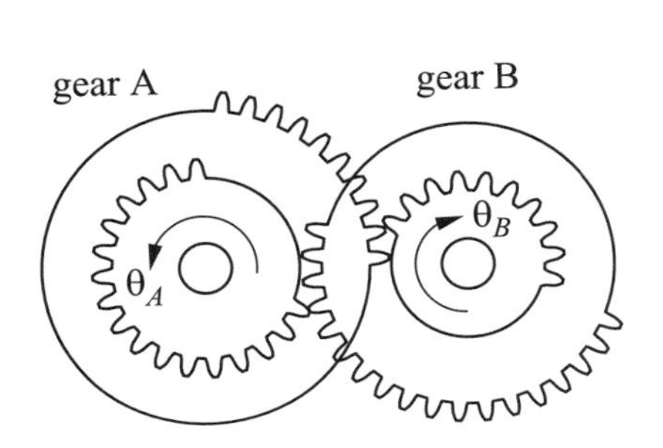

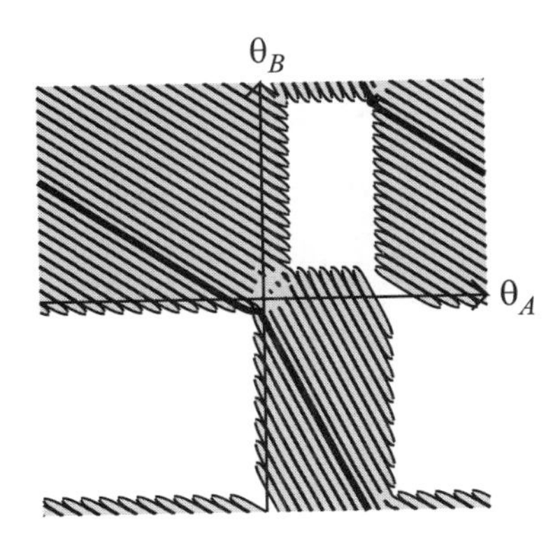

The pair consists of two rotating parts with orientations θ_A and θ_B. Each part consists of concentric large and small sector gears. Part A has a large gear sector with 10 teeth on top and a small gear sector with 17 teeth below. Part B has a small gear sector with 10 teeth on top and a large gear sector with 18 teeth below. The corresponding small and larger gear sectors on top and bottom alternatively engage. The pair has a structure similar to pair 17.

Rotating A causes B to rotate once in the opposite direction. The gear ratio is 2:3 when the lower gears segments mesh and is 2:1 when the upper gears segments mesh. The transmission is almost linear, with slight play and chatter. The pair is bidirectional.

The configuration space partition consists of two types of slanted narrow channels and two free space regions. The slanted channels correspond to the driving periods at each relative orientation in which the gears engage. The channel width corresponds to the play between the two gears. The different inclinations of the channels correspond the gear ratios of 2:1 and 2:3. The configuration space partition is the overlay of two shifted versions of the pair 17 partition.

20 Reciprocating Indexer Pair

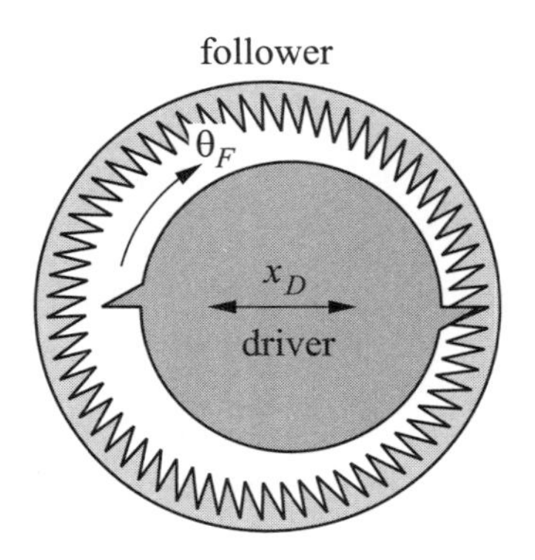

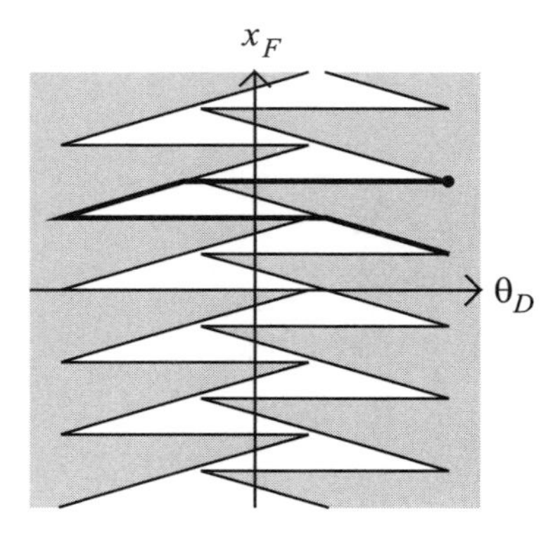

The pair consists of a reciprocating driver with position θ_D and a rotating follower with orientation θ_F. The driver has two triangular teeth slanted in opposite directions. The follower consists of an inner ring with 64 triangular teeth whose shape is complementary to that of the driver teeth. The driver teeth alternately engage the left and right follower slots.

As the drive moves left and right, it rotate the follower clockwise by one tooth per cycle. The pair is unidirectional.

The figure shows a detail of the configuration space partition. The boundaries of the left and right triangular-shape free regions correspond to the driver tooth contact with the follower teeth. The narrow passage between the two regions is the transition between the left and right contacts.

21 Spatial Indexer Pair

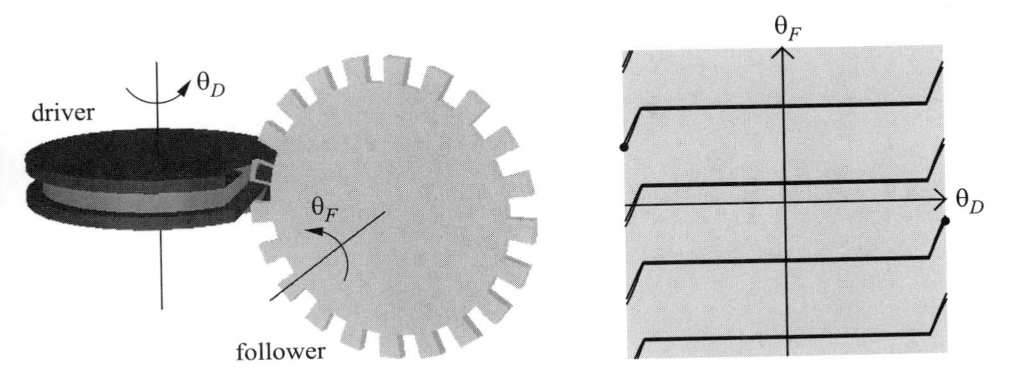

The pair consists of a rotating driver with orientation θ_D and a rotating follower with orientation θ_F. The driver consists of two cylindrical guides connected by a slanted crossover. The follower is a gear with 21 rectangular teeth. The rotation axes of the parts are perpendicular.

A full rotation of the driver advances the follower by one tooth. The crossover rotates the follower, then the guides lock it for the remainder of the driver's rotation. The pair is bidirectional.

The figure shows a detail of the configuration space partition; the full partition consists of 21 horizontal and slanted channels. The horizontal channels correspond to the dwell periods of the follower. The slanted channels correspond to the driving periods. The channel width is the play between the driver and the follower.

22 Spatial Geneva Pair

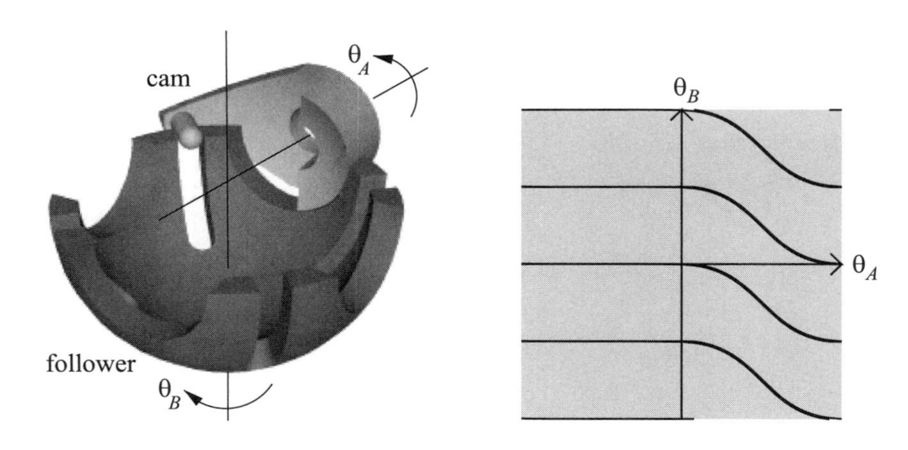

The pair consists of a rotating cam with orientation θ_A and a rotating follower with orientation θ_B. The cam is a plate with a cylindrical pin and a half cylinder mounted on it. The follower is a hollow hemisphere with four evenly spaced slots and circular cutouts. The follower's inner boundary consists of a spherical patch with four slots and four cutouts. As the cam rotates, the pin engages the follower slots and rotates it nonuniformly in the opposite direction. The cam cylinder engages a follower cutout during the idle periods, preventing its rotation.

As the driver rotates, the pin engages the Geneva wheel slots and rotates it nonuniformly in the opposite direction. The locking arc segments of the driver engage those of the wheel during the idle periods, thereby preventing the wheel from rotating. A full rotation of the driver causes a nonuniform, intermittent rotation of the wheel with four dwell periods and four drive periods. The pair is bidirectional.

The configuration space partition consists of single narrow channel whose nearly uniform vertical width is the wheel's backlash. The slanted channel segments correspond to the driving periods, whose boundaries are sinusoidal curves. The configuration space partition is nearly identical (with a 90° shift) to that of the planar Geneva pair 6.

23 Spatial Orthogonal Gear Pair

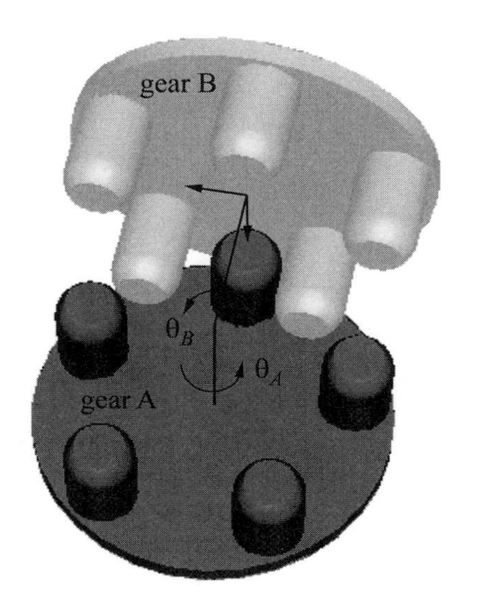

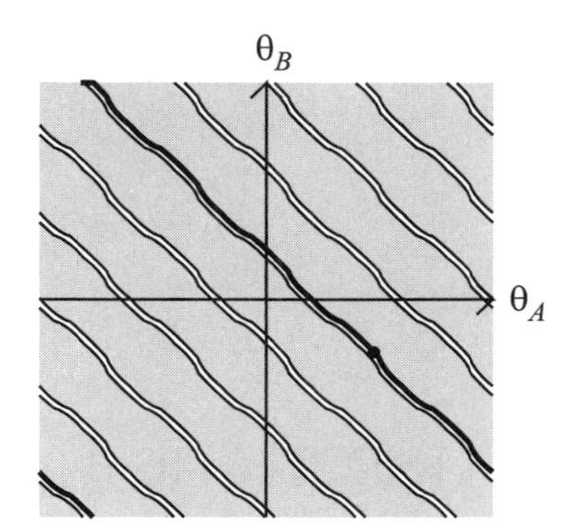

The pair consists of identical toothed gears rotating around orthogonal axes with orientations θ_A and θ_B. Each gear consists of five cylindrical teeth mounted on a cylindrical base at evenly spaced orientations.

When one gear rotates in either direction, the other gear rotates in the opposite direction by the same angle. Two pairs of teeth are engaged at all times. The transmission ratio is almost linear. The pair is bidirectional.

The configuration space consists of a single narrow channel that wraps around the vertical and horizontal axes. The channel width shows the amount of gear backlash. The channel boundaries are nearly linear. The rough edges reflect chatter and slight variations in the motion's transmission ratio. The configuration space partition is similar to that of the engaged (left) portion of the pair 17 partition.

24 Spatial Bevel Gears Mechanism

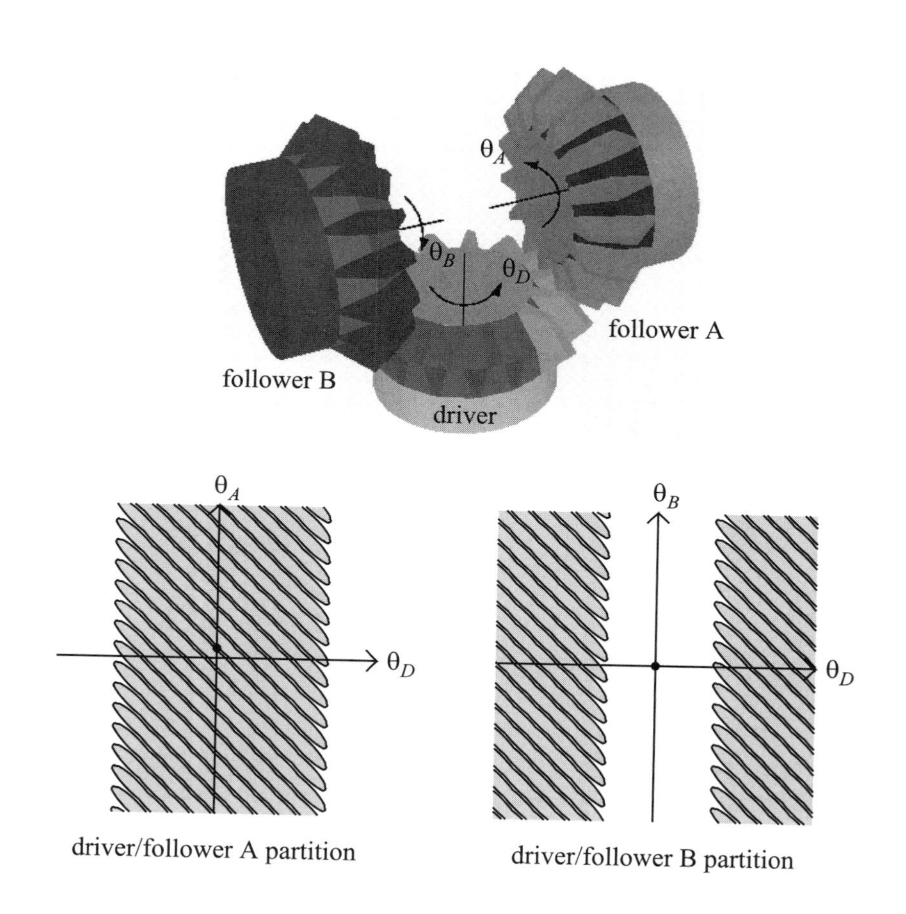

driver/follower A partition   driver/follower B partition

The mechanism consists of three rotating bevel gears: an 8-tooth driver gear with orientation θ_D and two 15-tooth follower gears with orientations θ_A and θ_B around a common rotation axis. The transmission ratio is almost linear, with slight play and chatter.

A rotation of the driver gear alternately engages the right and left gears and turns them by half a turn. When the driver gear engages one follower gear, the other can rotate freely. The driver gear is bidirectional.

The two driver-follower configuration space partitions consist of 15 slanted narrow channels and a free space region. They are identical with θ_D shifted by half a turn. The slanted channels correspond to the driving periods at each of the 15 possible relative orientations in which the gears engage. The channel width is the play between the two gears. The free space corresponds to independent gear rotations. The configuration space partitions are identical to that of pair 17.

25 Counter Mechanism

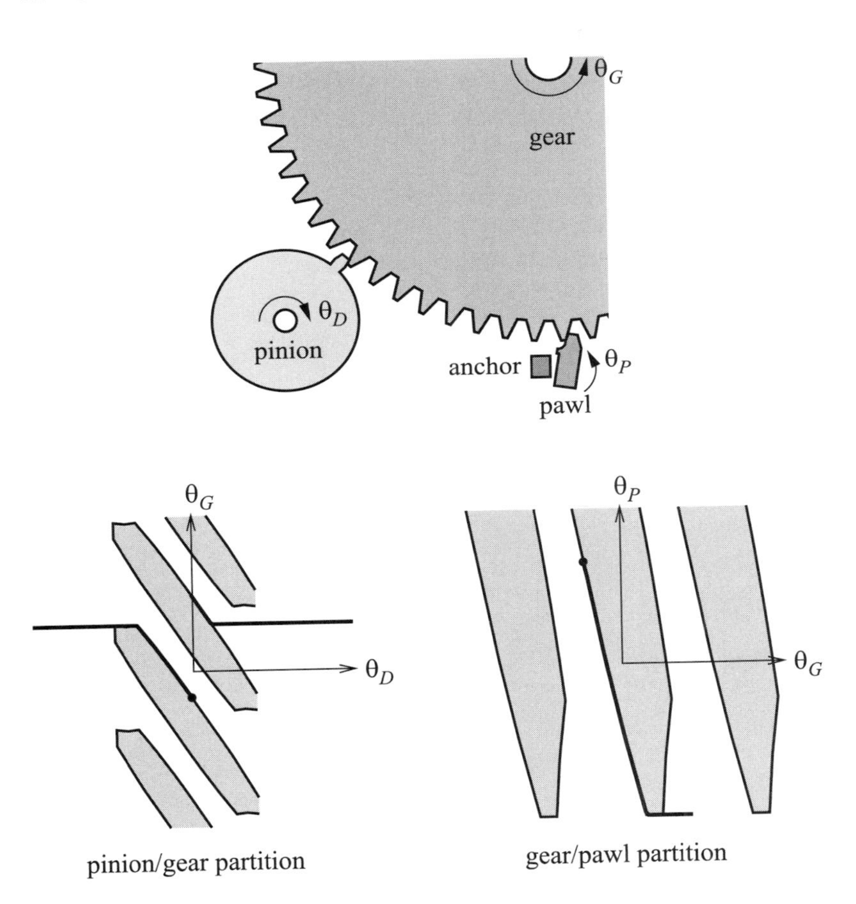

pinion/gear partition

gear/pawl partition

The mechanism consists of a rotating pinion with orientation θ_D, a rotating gear with orientation θ_G, a rotating pawl with orientation θ_P, and a fixed anchor. The pinion consists of a disk and a single tooth. The gear consists of 72 trapezoidal teeth. The pawl consists of a finger and a notch on its left side. A torsional spring (not shown) rotates the pawl counterclockwise. The anchor is a square that prevents the counterclockwise rotation of the pawl.

The pinion rotates clockwise and advances the gear by one tooth per rotation. Reverse rotation is prevented by the pawl blocking against the anchor. Forward rotation is limited to one tooth per cycle.

The figure shows details of the configuration space partitions of the two pairs. Both partitions consist of slanted elongated strips forming narrow channels. The kinematic simulation consists of horizontal segments in free space where the pinion rotates and the gear is idle, and of a slanted segment where the pinion rotates the gear clockwise by one-seventy-second turn, which in turn rotates the pawl clockwise.

26 Pawl Indexer Mechanism

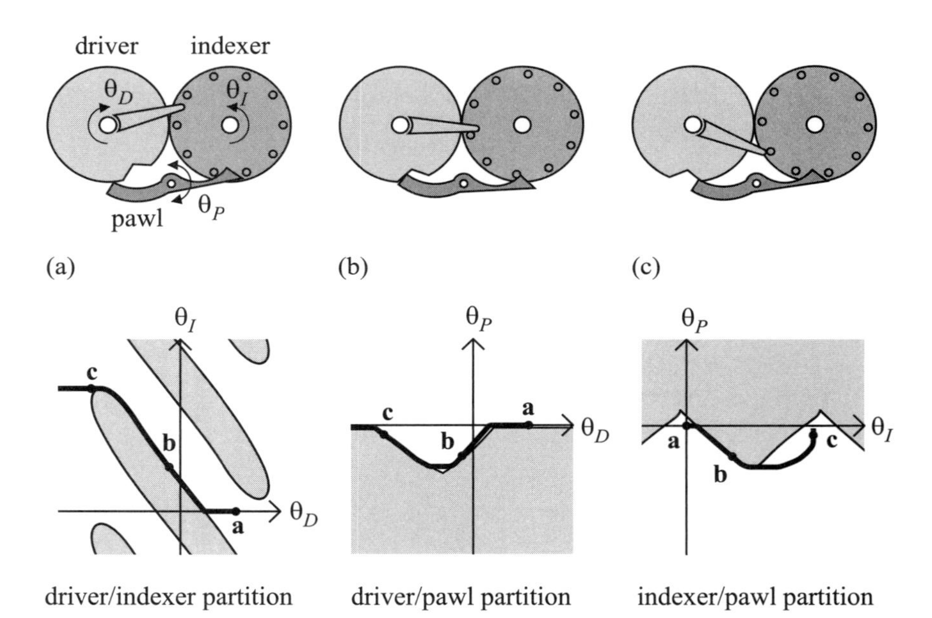

The mechanism consists of a rotating driver with orientation θ_D, a rotating indexer with orientation θ_I, and a rotating pawl with orientation θ_P. The driver consists of a disk with a notch and a finger. The indexer consists of a disk with ten evenly spaced circular pins. The pawl consists of left and right tips. The left tip engages the driver notch and the right tip engages the indexer pins.

As the driver rotates clockwise, its finger engages an indexer pin, rotates the indexer counterclockwise by one-tenth of a turn, and disengages. When the driver is disengaged (a), its rim aligns with the left pawl arm, which prevents clockwise rotation, and the right pawl arm prevents indexer rotation by engaging two pins. When the driver engages the indexer, its notch aligns with the left pawl arm, which allows the bottom indexer pin to rotate the right pawl arm out of the way (b). When it disengages, the left pawl arm returns to its rim and the right arm re-locks the indexer (c). The mechanism is unidirectional.

The driver-indexer free space consists of ten diagonal channels where the finger engages the ten pins. The negative slope indicates that the parts rotate in opposite directions. The rest of free space consists of the regions to the left and right of the channels where the driver is disengaged. The driver-pawl free space is bounded from below by the rim-left arm contact curve, which is horizontal, and by the notch-left arm contact curves, which form a v-shape. The horizontal curve blocks downward vertical motion, which corresponds to clockwise pawl rotation, whereas the slanted curves permit it. The indexer-pawl free space is bounded from above by the pin-right arm contact curves, which form ten inverted v-shapes. When θ_I increases or decreases, θ_P decreases along one side of the v-shape. Thus, indexer rotation in either direction causes clockwise rotation of the pawl.

27 Intermittent Gear Mechanism

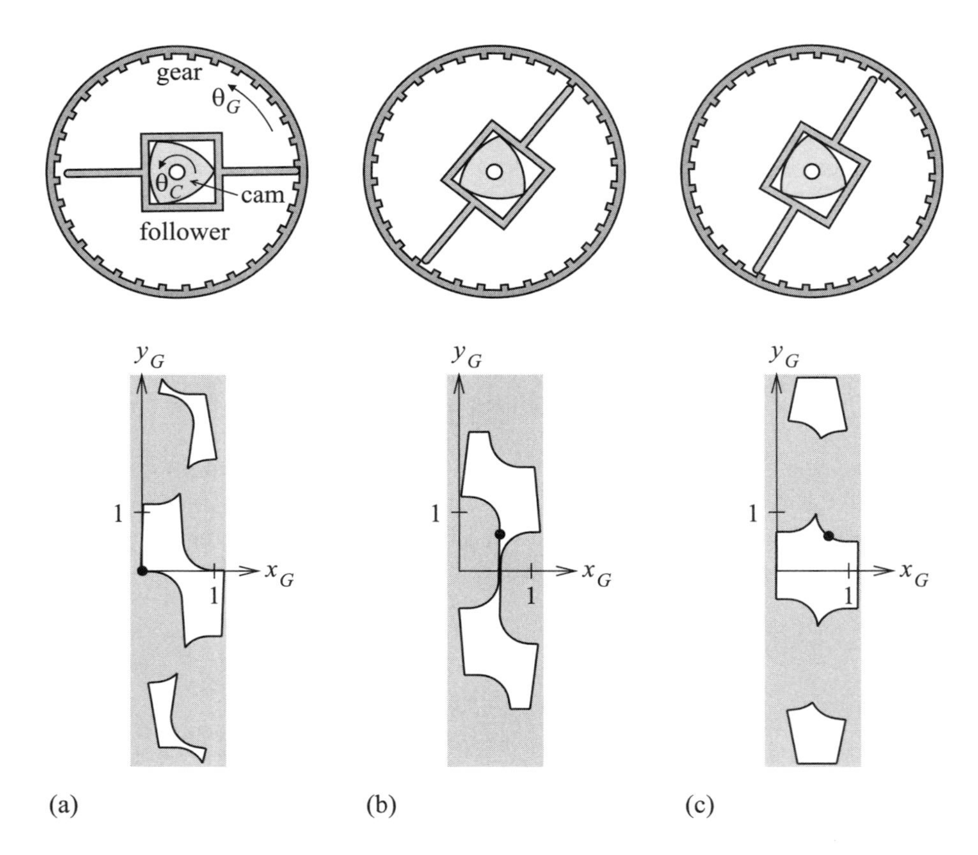

The mechanism consists of a constant-breadth cam, a square follower with two pawls, and a gear with inner square teeth. The cam and the gear are mounted on a fixed frame and rotate around their centers. The follower is free and has thus three degrees of freedom. The mechanism has two general planar pairs: cam-follower and gear-follower.

Rotating the cam causes the follower to rotate in step while reciprocating along its length. The right follower pawl engages a gear tooth at cam angle $\theta_C = 0°$ (a), the follower rotates the gear 45°, the right pawl disengages at $\theta_C = 82°$ (b), the follower rotates independently while the gear dwells, then the left pawl engages the gear at $\theta_C = 90°$ and the cycle repeats (c).

The bottom row of figures shows details of the cross-sections of the gear-follower configuration space partitions at the gear angles of the top figures. The follower configuration is $(0,0,0)$ and the gear degrees of freedom are (x_G, y_G). In (a), free space consists of three regions. In the middle region, which contains the snapshot configuration, the right follower pawl engages a gear tooth. In the region above and below, the pawl engages the tooth below and above, respectively. In (b), the three regions have merged into a single region, which shows that the follower can now cross over the two gear teeth. In (c), free space once again consists of three regions and the middle region contains the snapshot configuration in which the left pawl engages a gear tooth.

28 Lever Indexer Mechanism

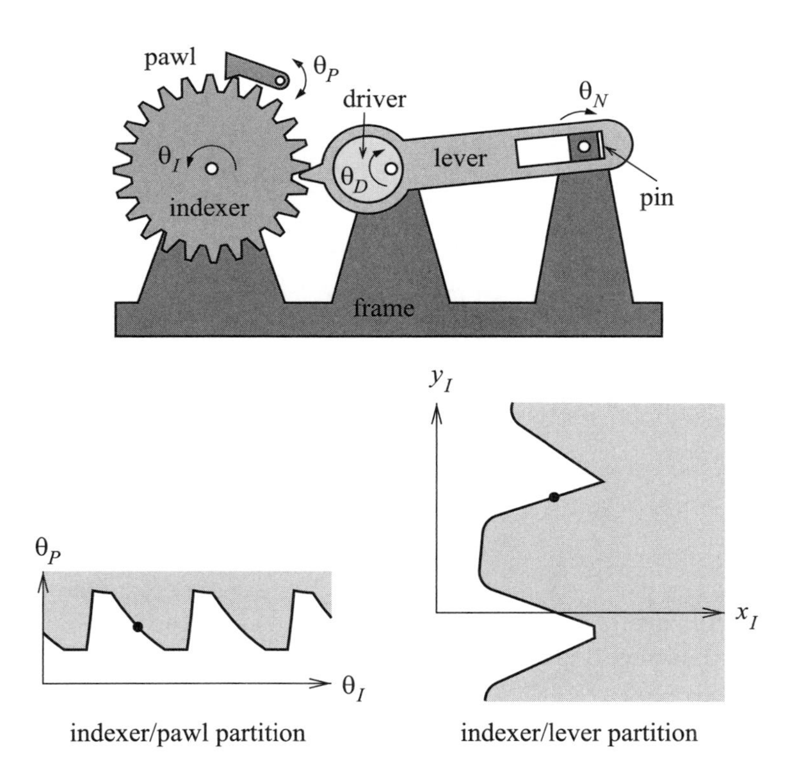

The mechanism consists of a rotating driver with orientation θ_D, a rotating indexer with orientation θ_I, a rotating pawl with orientation θ_P, a rotating pin with orientation θ_N, a free lever with degrees of freedom (x_L, y_L, θ_L), and a frame. The driver is an offcenter cylinder that acts as a cam. The indexer is a gear with 24 trapezoidal teeth. The pawl has a triangular tip shaped to follow the indexer teeth. The pin is a rectangular block. The lever has a rounded triangular tip shaped to engage the indexer teeth, a cylindrical hole that fits over the cam, and a rectangular slot that fits over the pin.

The mechanism advances the indexer wheel by one tooth (15°) for every turn of the driver. As the driver turns clockwise, the lever tip traces a closed trajectory whose form is determined by the relative position of the driver and pin rotation axes, the driver's offset, and the length of the lever. This causes the indexer to rotate counterclockwise by 15°. The pawl prevents clockwise rotation of the indexer. The mechanism is unidirectional.

The indexer-pawl partition detail shows that the vertical boundary segment is what prevents clockwise rotation of the indexer. The pair 14 partition is similar. The right-hand figure is a detail of the cross-section of the indexer-lever partition.

29 Movie Film Advance Mechanism

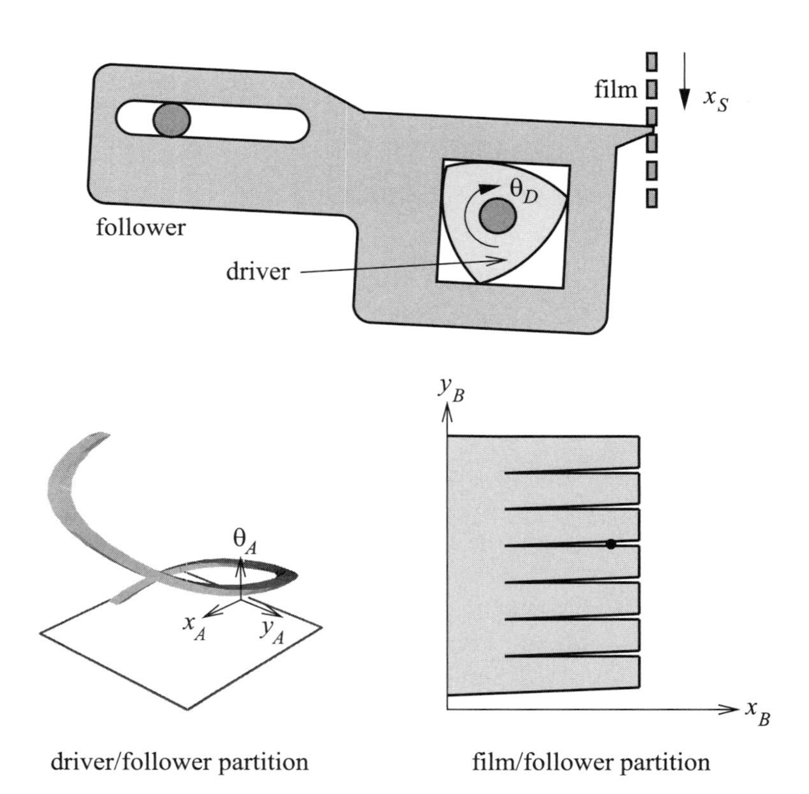

driver/follower partition   film/follower partition

The mechanism consists of a rotating driver with orientation θ_D, a follower with degrees of freedom (x_F, y_F, θ_F), and a vertically translating film strip with position y_S. The driver is a constant-breadth cam. The follower consists of a horizontal slot, a square in which the driver cam fits, and a finger that engages the film notches. The follower is attached to the frame by a pin joint. The film is orthogonal to the figure's plane. Its cross-section consists of notches.

As the driver rotates counterclockwise, the follower tip engages the film, pushes it down one frame, and retracts. The mechanism is bidirectional.

The coordinates of the driver-follower configuration space partition are (x_A, y_A, θ_A). The free space is a narrow spiral channel. The coordinates of the film-follower configuration space partition are (x_B, y_B, θ_B). The figure shows a $\theta_B = 0$ cross-section detail. The other cross-sections are similar, with rotated engagement curves.

30 Camera Shutter Mechanism

Figure A.1a–b shows the shutter mechanism of a 35-mm film camera. It consists of a rotating driver with orientation θ_D, a rotating shutter with orientation θ_S, a rotating shutter lock with orientation θ_L, a rotating advance wheel with orientation θ_W, a rotating pawl with orientation θ_P, a rotating film counter (not shown) with orientation θ_C, and a horizontally translating film with position x_F. The driver consists of a cam, a slotted wheel, a film wheel, and a counter wheel mounted on a shaft. The shutter consists of a tip and a pin, and is spring loaded counterclockwise. The shutter lock is spring loaded clockwise. Figures A.2a,c,e and A.3a,c,e show details of six pairs.

The mechanism alternately loads and exposes the film frames. In the loading mode, the user turns the wheel, which advances the film, which rotates the driver. Loading ends when the driver engages the shutter in the lock. In the exposure mode, the user presses the release button, which rotates the lock, which releases the shutter. The shutter spring rotates the shutter, which trips the curtain, which rotates away from the lens and exposes the film.

Figure A.2b,d,f shows details of configuration space partitions for three driver pairs. The boundary of the driver-shutter tip configuration space partition shows how the driver cam raises and lowers the shutter tip (figure A.2b). The configuration space partition for the driver-shutter lock shows how the shutter lock tip prevents the driver wheel from rotating when it is inside its slot (figure A.2d). The configuration space partition for the

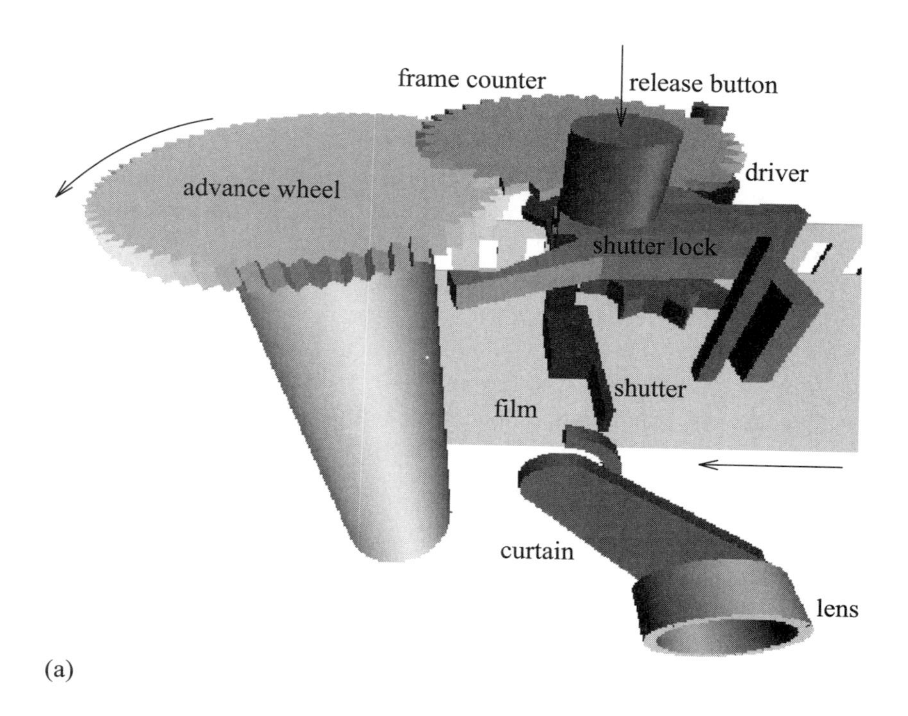

(a)

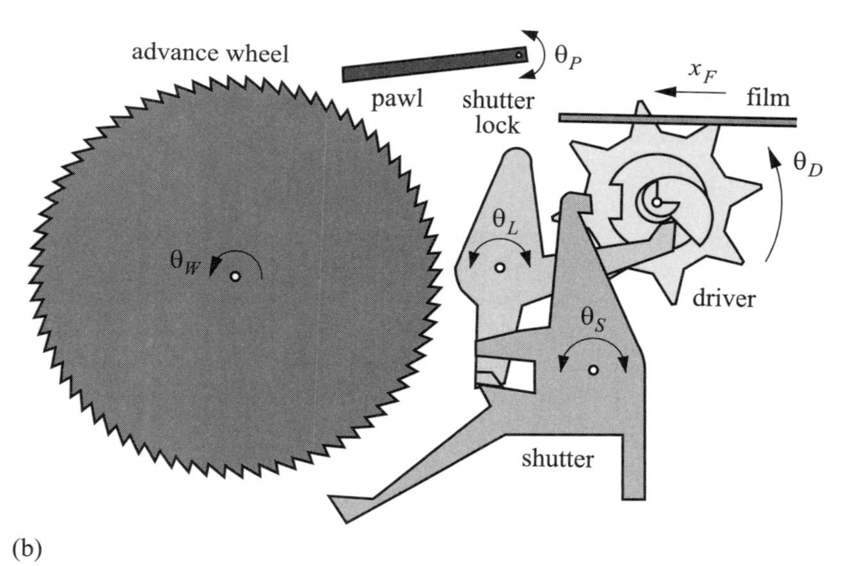

(b)

Figure A.1
Camera shutter mechanism: (a) 3D view, (b) 2D view.

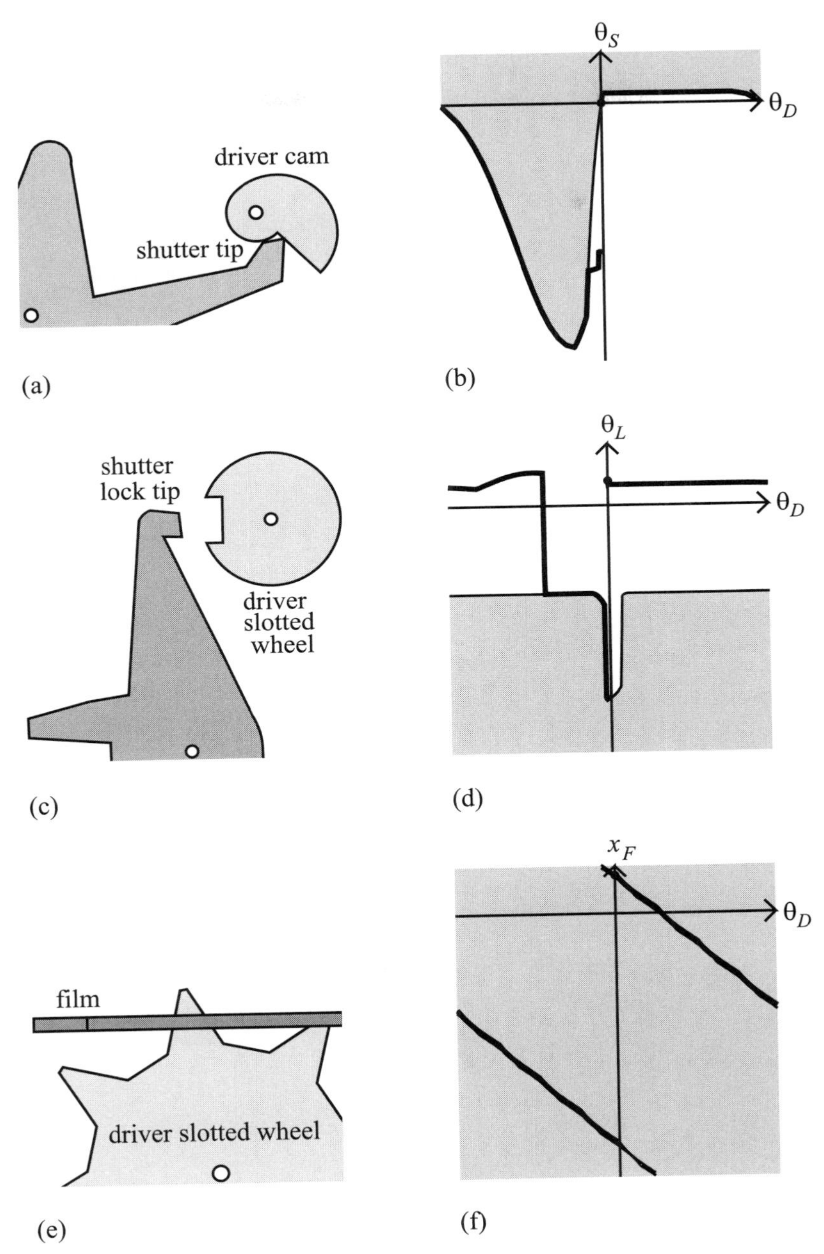

Figure A.2
Pairs and details of their configuration space partitions: (a–b) driver-shutter tip, (c–d) driver-shutter lock, (e–f) driver-film.

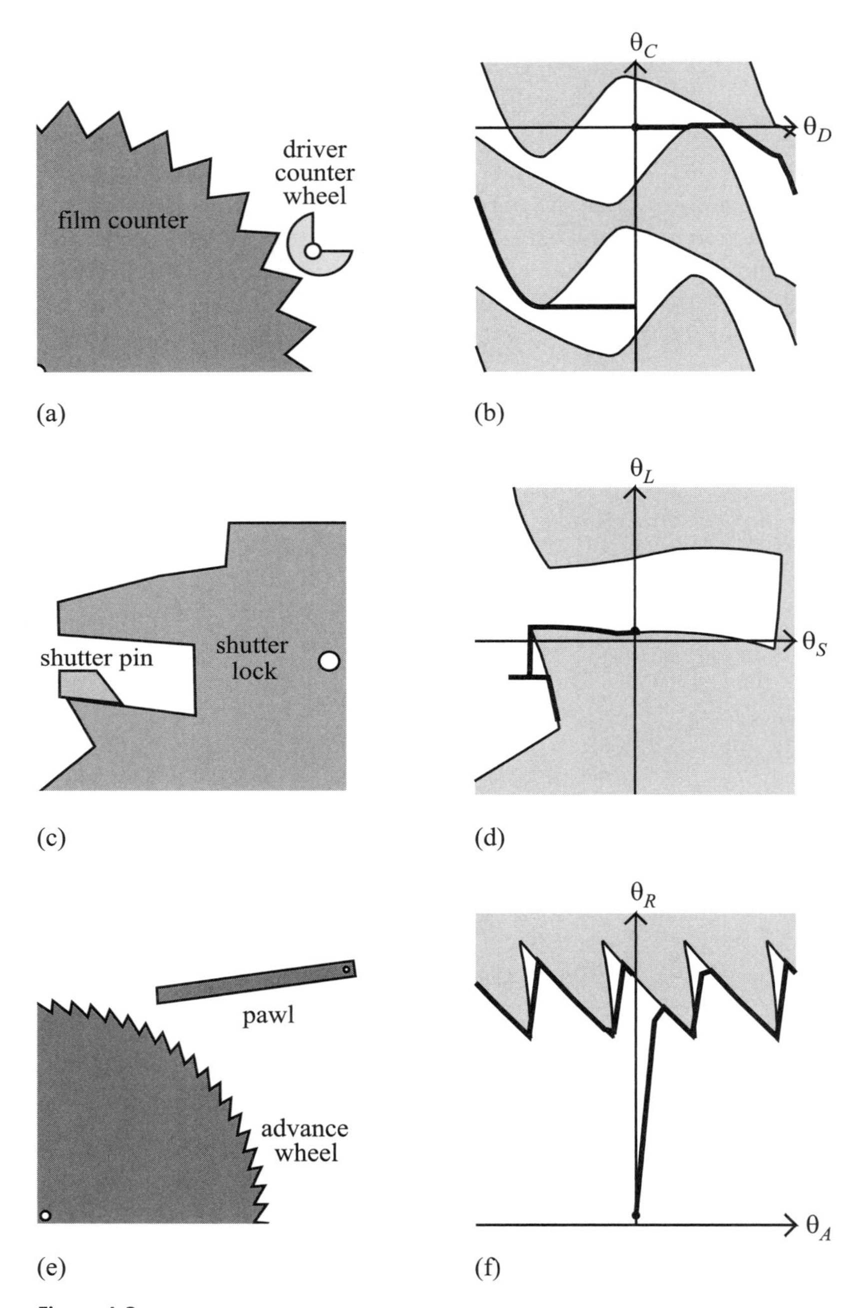

Figure A.3

Pairs and details of their configuration space partitions: (a–b) driver-film counter, (c–d) shutter lock-shutter pin, (e–f) advance wheel-pawl.

driver-film is a single narrow channel that shows the nearly linear relation between the driver's rotation and the film advance with almost no play.

Figure A.3b,d,f shows details of configuration space partitions for three driver pairs. The configuration space partition for the driver counter-film counter consists of narrow channels at each tooth orientation. Note that the driver counter locks the film counter when not advancing it (figure A.3b). The notch in the configuration space partition for the shutter lock-shutter pin (figure A.2d) shows how the shutter lock locks and releases the shutter pin. The configuration space partition for the advance wheel-pawl (figure A.2f) is similar to the configuration space partition for pair 14.

Appendix B: HIPAIR Software

In this appendix, we describe the HIPAIR mechanism design program built around computation, visualization, and manipulation of the configuration space. We describe installation, the graphical user interface (GUI), and the input format. The program is written in Ansi C++ using openGL and glut. It is free for education and research. Send comments, bug reports, and request for commercial licenses to Elisha Sacks (eps@cs.purdue.edu).

B.1 Installation

The website is `http://www.cs.purdue.edu/archives/2008/eps/`. The program runs under Linux and Windows. The archive `hipair.tar` contains the source files, which are the same for both operating systems, and the `ex` subdirectory, which contains input files for most of the mechanisms in appendix A. The file `hipair.exe` is a windows executable.

Here is how to install and run the Linux version.

```
% tar -xvf hipair.tar

% make

% hipair ex/3finger
```

If the argument is omitted, HIPAIR prompts for it and reads it from standard input. In Windows, create a project with the source file suffix changed to .cpp and with glut installed.

B.2 Graphical User Interface

The program opens a parts windows and one cspace (configuration space) window per pair (figure B.1). The windows can be resized. The parts window shows the parts in the current configuration. Each cspace window shows its configuration space with the animation path in red, the

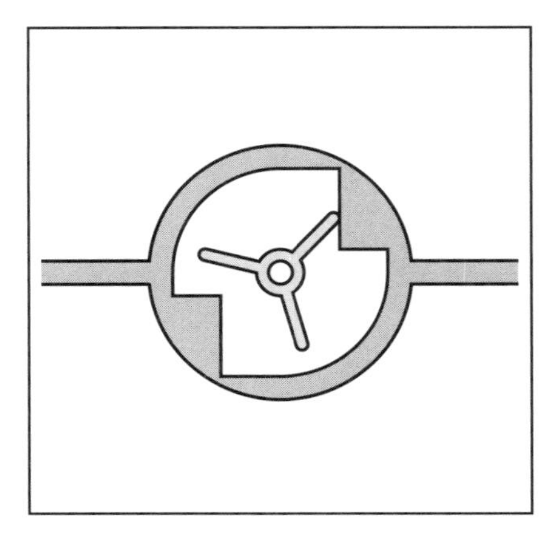
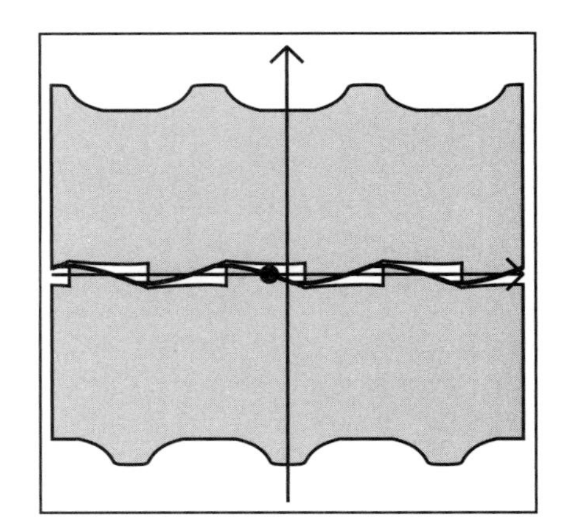

Figure B.1
Parts window (a) and cspace window (b) for ex/3finger.

current configuration as a green dot, and the contact zone in green when requested. The program normally displays the current animation step. The user can increase or decrease the step, or can animate the mechanism. The animation starts at the current step and ends at the last step or when interrupted. The animation speed is specified as a multiple of one-thirtieth of a second. The user can specify an arbitrary configuration with the mouse. The next animation request resets the configuration to the current animation step. The user can set the contact zone width and can toggle the contact zone display.

The program is controlled by the mouse. Press, drag, and release the left button to zoom to a rectangle. Click the middle button in a cspace window to set its two configuration parameters to the mouse's position. Click the right button to bring up this menu. The menu entries can also be invoked by typing the letters in parentheses.

- Start/stop animation (a).
- Next animation step (n).
- Previous animation step (p).
- Set animation speed (s).
- Draw parts in wireframe/filled (f).
- Restore initial viewing box (r).
- Show/hide contact zones (z).
- Set contact zone width (w).
- Save window in postscript (o).
- Exit (q).

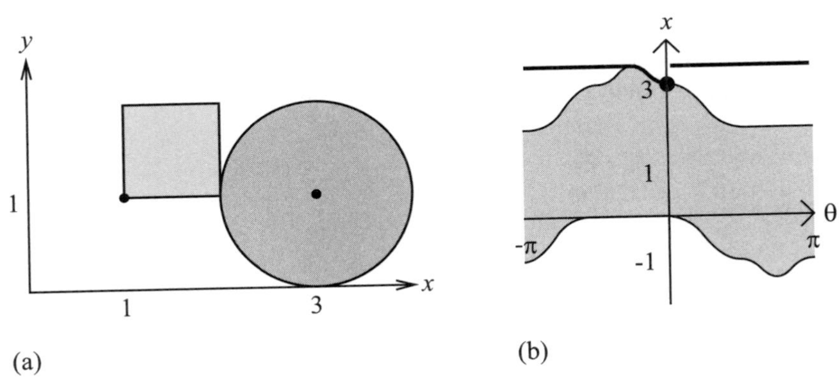

Figure B.2
Square-circle pair (a) and its configuration space partition (b).

B.3 Mechanisms

A mechanism consists of planar parts that form higher kinematic pairs. HIPAIR computes configuration spaces for the pairs, computes the part's motions for given driving motions, and computes contact zones that bound the kinematic variation for the given part tolerances. This section explains these concepts.

B.3.1 Parts

A part is specified in a parts coordinate system. The part consists of regions in the *xy* plane that are extruded over intervals on the *z* axis. A region has an outer boundary and may also have inner boundaries. A boundary is a simple loop of line and circle segments.

For example, a cylinder of 10 mm height is specified as a circle in the *xy* plane that is extruded over $0 \leq z \leq 10$. The region boundaries of a slice must be disjoint, but slices may overlap. A part has an initial position and orientation in world *xy* coordinates. It has one degree of freedom: rotation around its origin, which is fixed at its initial position, or translation along a line that passes through its initial position.

Figure B.2a shows a simple mechanism composed of two parts. Each part consists of a single slice with *z* range $[0, 1]$. The first part is a unit square. Its origin is the bottom left corner. Its boundary consists of four line segments. The part rotates with initial position $(1, 1)$ and orientation 0. The second part is a circle of radius 1 whose origin is the circle's center. The part translates horizontally with initial position $(3, 1)$ and orientation 0.

B.3.2 Configuration Spaces of Fixed-Axis Pairs

Interactions between pairs of parts are modeled as configuration spaces. The configuration space is a two-dimensional manifold whose coordinates are the parts' degrees of freedom. Figure B.2b shows the configuration space of the square-circle pair. It is a cylinder whose coordinates are the square orientation θ and the rectangle offset x. The configuration space is a torus when both parts rotate and is a rectangle when both translate. Configuration space partitions into free space where the parts do not touch (the white region in the figure), blocked space where they overlap (the gray region), and contact space where they touch (the black curves). The large dot marks the displayed configuration of $(0, 3)$.

B.3.3 Kinematic Simulation

Kinematic simulation takes the driving motion of a part as input and computes motions for the other parts that prevent overlap. In our example, the driving motion is clockwise rotation of the square by one full turn. The circle translates clear of the square, then stands still. The motion's path is the thick black curve in figure B.2b.

B.3.4 Contact Zones of Fixed-Axis Pairs

Variation in the shapes of parts is modeled as a zone around the nominal boundaries of the part where the actual boundaries are allowed to lie. HIPAIR supports constant-width offsets. As the boundaries of the parts in a kinematic pair range over their zones, their contact spaces vary in a zone around the nominal space, called the contact zone of the pair. The contact zone topology and geometry model the kinematic variation that is due to the parts' variation. Figure B.3 shows the square-circle contact zone for offsets of 0.1 on both parts. It forms a narrow band around the nominal con-

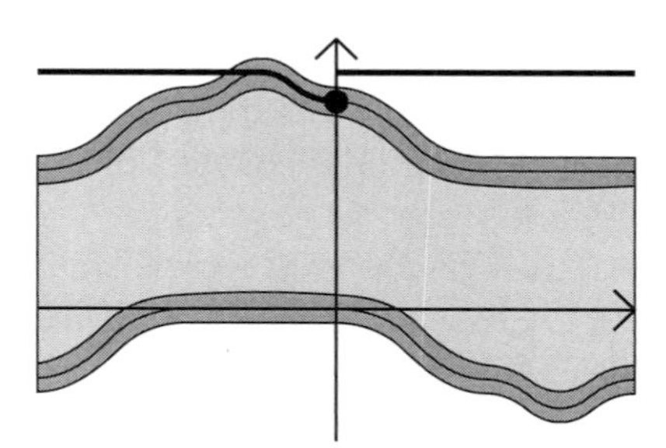

Figure B.3
Square-circle contact zone.

tact space, which shows that variation in the parts merely perturbs the nominal kinematic function.

B.4 Input Format

The input is a mechanism, simulation data, and display data.

B.4.1 Mechanisms

The mechanism format is specified by the following BNF formulas. The first line specifies the number of parts and pairs, then come the parts, then the pairs. The parts are assigned indices starting from zero. A pair is specified by two part indices and the accuracy of the configuration space computation.

MECHANISM	=	1 nparts npairs 0 0 PART* PAIR*
PART	=	name nslices color parameters SLICE+
SLICE	=	nboundaries zlower zupper BOUNDARY+
BOUNDARY	=	ngroups SGROUP+
SGROUP	=	nsegments SEGMENT+ noffsets offset*
SEGMENT	=	LINE_SEGMENT or ARC_SEGMENT
LINE_SEGMENT	=	index 0 tx ty hx hy flag
ARC_SEGMENT	=	index 1 tx ty hx hy flag cx cy r s e
PAIR	=	index1 index2 accuracy

B.4.2 Parts

A part is specified by a name, number of slices, color, parameters, and slices. The name is a character string. The color is three integers in the range $[0, 255]$ followed by a dummy integer. The parameters are seven doubles. Parameter 1 is the motion type: 1 for translation along the x axis, 2 for translation along the y axis, 3 for rotation, and 4 for translation along a slanted axis. Parameters 2–4 are the initial position and orientation of the part. Parameters 5–6 are the lower and upper bounds of the part's motion parameter. For motion type 4, parameter 7 is the slope of the translation axis. The motion parameter and its bounds are measured along this axis.

A slice is specified by the number of boundaries, the lower and upper z values, the outer boundary, and the inner boundaries. A boundary is specified by the number of segment groups followed by the groups. A group is a sequence of incident segments and a sequence of offsets. A segment is specified by a dummy index, 0 or 1 for a line or circle, a tail (t_x, t_y), a head

```
1 1 2 1 0 0

square 1 255 0 0 0
3.0 1.0 1.0 0.0 -3.14159 3.14159 0.0
1 0.0 1.0
1 4
0 0 0.0 0.0 1.0 0.0 1
0 0 1.0 0.0 1.0 1.0 1
0 0 1.0 1.0 0.0 1.0 1
0 0 0.0 1.0 0.0 0.0 1
1 0.0

circle 1 0 255 0 0
1.0 4.0 1.0 0.0 -5.0 5.0 0.0
1 0.0 1.0
1 2
0 1 -1.0 0.0 1.0 0.0 1 0.0 0.0 1.0 -3.14159 0.0
0 1 1.0 0.0 -1.0 0.0 1 0.0 0.0 1.0  0.0 3.14159
1 0.0

0 1 0.001
```

Figure B.4
Square-circle file format.

(h_x, h_y), and a flag that is 1 when the part's interior lies to the left when the segment is traversed from tail to head. A circle segment also has a center (c_x, c_y), radius r, start angle s, and end angle e.

The segments in the group are repeated *noffset* times and the ith copy is offset by the ith element of the offset sequence. The offset is along the part's motion axis: the segments of a rotating part are rotated and the segments of a translating part are translated. Offsets are optional since any boundary can be specified as one group with noffset 1 and offset 0.

Figure B.4 illustrates the file format with the square-circle mechanism from figure B.2. The first line indicates two parts and one pair. Lines 3–11 describe the first part. Line 3: name "square," one slice, and red green blue (RGB) color $(255, 0, 0)$. Line 4: motion type 3 (rotation), initial position $(1.0, 1.0)$, initial orientation 0.0 radians, and rotation range $[-\pi, \pi]$ radians. Line 5: the slice has one boundary and has z extent $[0.0, 1.0]$. Line 6: the boundary consists of one sgroup with four segments. Lines 7–10: four line segments. Line 11: trivial shift sequence, explained next. Lines 13–20 specify the circle: name "circle," one slice, RGB color $(0, 255, 0)$, motion type 1

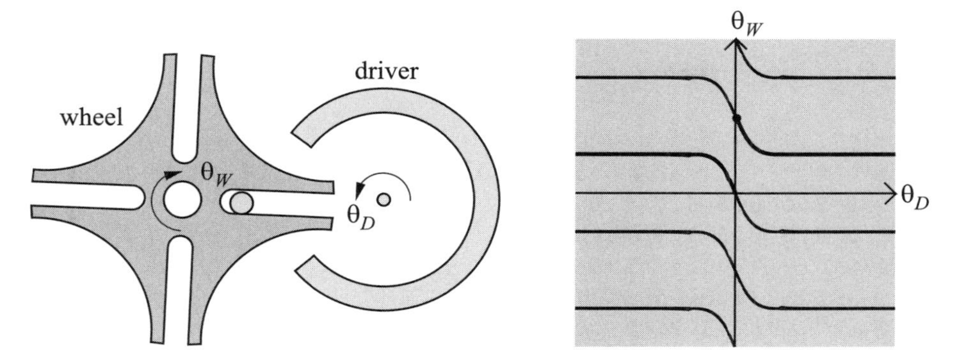

Figure B.5
Geneva pair (a) and its configuration space (b).

(x translation), initial position $(1.0, 4.0)$, initial orientation 0.0, translation range $[-5, 5]$, the slice has one boundary with two segments, two circle segments, and the shift sequence. The last line specifies the one pair: part index 0 (the square), part index 1 (the circle), and configuration space accuracy 0.001.

Figure B.5 shows a Geneva mechanism. The configuration space is a torus because both parts rotate. The driver has three slices with z interval $[0, 1]$: a locking arc, a driving pin, and a central pin. The three slices are connected by a plate (not shown) with z interval $[-1, 0]$. The wheel has one slice. The outer boundary consists of a six-segment sgroup with four offsets. The inner boundary is one circle. Figure B.6 shows the follower specifications. The offsets, 0, $\pi/2$, π, $3\pi/2$ radians, generate four rotated copies of the six segments.

B.4.3 Simulation and Display Data

The simulation data are specified by the following BNF formulas.

MOTIONS	=	nm MOT_DESCR* CONFIG
MOT_DESCR	=	mtime nm_parts PART_MOT_DESCR*
PART_MOT_DESCR	=	pindex velocity
CONFIG	=	one double per part

The number of motions is followed by one descriptor per motion. A motion descriptor is the motion's time, the number of moving parts, then the descriptors of the part's motion. A part's motion descriptor is a part index followed by a velocity. The motion begins with the part's motion

```
follower 1 0 255 0 0
3.0 0.0 0.0 0.0 -3.14159 3.14159 0.0
2 0.0 1.0
1 6
0 0 6.0 -0.5 2.0 -0.5 1
0 1 2.0 0.5 2.0 -0.5 0 2.0 0.0 0.5 1.5708 -1.5708
0 0 2.0 0.5 6.0 0.5 1
0 0 6.0 0.5 5.9585 0.9983 1
0 1 0.9983 5.9585 5.9585 0.9983 0 5.6568
  5.6568 4.6683 3.07693 -1.5061
0 0 0.9982 5.9585 0.5 6.0 1
4 0.0 1.5708 3.14153 4.712389
1 1
0 1 -0.75 0.0 -0.75 0.0 0 0.0 0.0 0.75
  -3.14159 3.14159
1 0.0
```

Figure B.6
Follower file format.

parameters specified in CONFIGURATION. In the square-circle example, the motion sequence of rotating the square with angular velocity -1.0 for 7.0 seconds, then translating the circle with x velocity 3.0 for 5.0 seconds has the following file descriptor. The starting configuration is $\theta = 0$ and $x = 3$.

```
2

7.0 1 0 -1.0

5.0 1 1 3.0

0 3
```

The display data are two dummy values followed by a part bounding box. It is

```
0 0 -5 5 -5 5
```

in our example.

Glossary

axis a fixed line in space.

boundary representation representation of a part's shape describing the shape in terms of its features.

conceptual design the task of selecting a design concept that captures the desired kinematic function.

configuration position and orientation of one coordinate frame with respect to another.

configuration space vector space that specifies all possible configurations for a part or a mechanism.

configuration space, blocked space subspace of configuration space where parts overlap.

configuration space, cross-section cross-section of a general planar pair configuration space in which the orientation of the moving part is fixed.

configuration space, contact space subspace of configuration space where parts are in contact.

configuration space, free space subspace of configuration space where parts are disjoint.

configuration space partition decomposition of a configuration space into free, blocked, and contact spaces.

configuration space, region subspace of configuration space.

configuration space, topology configuration space regions and their adjacency relations.

configuration variable position and orientation variables.

contact parts whose interiors are disjoint and whose boundaries overlap.

contact change transition from one pairwise feature contact to another.

contact configuration configuration in which parts are in contact.

contact curve zero set of contact equation in two variables.

contact equations equation in the configuration variables of two parts whose zero set consists of contact configurations.

contact type contact between two features of a part. Contact types between simple planar features include circle-circle, circle-line, line-line, vertex-vertex, circle-vertex, and line-vertex contacts.

contact zone union of the contact spaces of the mechanisms in a tolerance space.

cross-section two-dimensional slice of a part.

degrees of freedom independent configuration variables that are needed to specify the configuration of a part or mechanism.

design space hyperbox in a configuration space that is the cross-product of one design interval per parameter.

dynamical simulation computation of the motions of parts resulting from the input and external forces acting on the parts.

feature element of a part's boundary. Simple features include points and line and arc segments in the plane and vertex, edges, or faces in space.

fixed-axis mechanism mechanism whose parts move along axes that are fixed in space.

fixed-axis motion rotation or translation along an axis that is fixed in space.

fixed-axis pair kinematic pair composed of fixed-axis parts.

fixed-axis part part with one degree of freedom (rotation or translation).

general planar pair two parts whose relative motions are circumscribed to a plane.

higher pair two parts not in permanent surface contact that move relative to each other.

HIPAIR open-source C++ software package that implements some of the kinematic design methods for mechanisms described in this book.

kinematic analysis derivation of the kinematic function of a mechanism from specification of the shapes of its parts and their motion constraints.

kinematic design process of designing a mechanism to achieve a desired kinematic function.

kinematic function function describing the transformation of input motions into output motions by the contact of parts.

kinematic pair two parts whose motions are constrained by contacts.

kinematic simulation computation of compliant motions of parts resulting from an input driving motion.

kinematic synthesis creation or modification of parts' shapes and motion constraints to achieve a desired kinematic function. The inverse of kinematic analysis.

kinematic tolerancing derivation and modification of the kinematic variation of a mechanism that is due to the shape of its parts and variations in configuration.

kinematic variation changes in the kinematic function of a mechanism that are due to variations in the shapes and configuration of its parts.

kinematics branch of mechanics that studies the motions of parts independently of the forces acting on them.

linkage mechanism formed by lower pairs only.

lower pair two parts in permanent surface contact that move relative to each other.

mechanism assembly of moving and fixed parts.

mechanism classification classification of mechanisms according to their structural and functional characteristics.

mechanism configuration position and orientation of a mechanism's parts.

mechanism design the process of creating or modfying a mechanism to achieve a desired function.

motion constraint motion ranges and relations induced by contact of parts.

nominal model parametric model with parameter values that achieve the intended kinematic function in the absence of variation that is due to manufacturing.

nominal parameter values parameter values of the nominal model.

pair configuration position and orientation of two parts.

parameter real-valued variable.

parametric design search for values of parameters that achieve a design goal.

parametric model description of the shapes and configuration of a mechanism's parts in terms of their parameters.

parametric optimization search for values of parameters that maximize a design quality metric.

parametric synthesis synonym of parametric design.

parametric tolerancing search for values of parameters that minimize design variation.

part rigid solid with a fixed shape.

part configuration position and orientation of a part.

planar mechanism mechanism whose parts are planar.

planar motion motions occuring in a plane (one rotation, two translations).

planar part part with a fixed cross-section with respect to a plane.

subsumed contact contact prevented by other feature contacts.

tolerance interval of allowable values for a parameter.

tolerance analysis analysis of the variation in a design due to tolerances on the parameters of a parametric model.

tolerance, nominal value numerical value of a design parameter in the nominal design.

tolerance space hyperbox in parameter space that is the cross-product of one tolerance interval per parameter.

tolerance, worst-case computation of the maximal variation from the nominal kinematic function over the tolerance space.

tolerancing the task of assigning tolerances to the shapes and configurations of parts.

topology graph graph representation of the contacts of parts in a mechanism. Each part is represented by a node, and edges between nodes represent contacts.

[58] Michael McCarthy and Leo Joskowicz. Kinematic synthesis. In E. K. Antonsson and J. Cagan, editors, *Formal Engineering Design Synthesis*, pages 321–362. Cambridge University Press, 2001.

[59] Parviz E. Nikravesh. *Planar Multibody Dynamics*. CRC Press, Abingdon, England, 2007.

[60] Yaron Ostrovsky-Berman and Joskowicz. Tolerance envelopes of planar mechanical parts with parametric tolerances. *Computer-Aided Design*, 37(5):531–544, 2005.

[61] Panos Papalambros and Douglass Wilde. *Principles of Optimal Design*. Cambridge University Press, 1988.

[62] Richard P. Paul. *Robot Manipulators: Mathematics, Programming, and Control*. MIT Press, Cambridge, MA, 1981.

[63] Umberto Prisco and Giuseppe Giorleo. Overview of current CAT systems. *Integrated Computer-Aided Engineering*, 9(4):373–387, 2002.

[64] Rajan Ramaswamy. "Computer tools for preliminary parametric design." PhD thesis, Massachussetts Institute of Technology, Cambridge, MA, 1993.

[65] Charles F. Reinholtz, Sanjay G. Dhande, and George N. Sandor. Kinematic analysis of planar higher pair mechanisms. *Mechanism and Machine Theory*, 13:619–629, 1978.

[66] Aristides A. G. Requicha. Mathematical definition of tolerance specifications. *Manufacturing Review*, 6(4):269–274, 1993.

[67] Franz Reuleaux. *The Kinematics of Machinery: Outline of a Theory of Machines*. Dover Publications, New York, 1963.

[68] Elisha Sacks. Practical sliced configuration spaces for curved planar pairs. *International Journal of Robotics Research*, 18(1):59–63, 1999.

[69] Elisha Sacks. Path planning for planar articulated robots using configuration spaces and compliant motion. *IEEE Transactions on Robotics and Automation*, 19(3), 2003.

[70] Elisha Sacks and Steven M. Barnes. Computer-aided kinematic design of a torsional ratcheting actuator. In *Proc. of the Fourth Int. Conference on Modeling and Simulation of Microsystems*. Nara Sciences and Technology Institute, Hilton Head, SC, 2001.

[71] Elisha Sacks and Leo Joskowicz. Computational kinematic analysis of higher pairs with multiple contacts. *Journal of Mechanical Design*, 117(2(A)):269–277, 1995.

[72] Elisha Sacks and Leo Joskowicz. Parametric kinematic tolerance analysis of planar mechanisms. *Computer-Aided Design*, 29(5):333–342, 1997.

[73] Elisha Sacks and Leo Joskowicz. Parametric kinematic tolerance analysis of general planar systems. *Computer-Aided Design*, 30(9):707–714, 1998.

[74] Elisha Sacks, Leo Joskowicz, Ralf Schultheiss, and Min-Ho Kyung. Towards robust kinematic synthesis of mechanical systems. In P. Bourdet and L. Mathieu, editors, *Geometric Product Specification and Verification: Integration of Functionality*, pages 135–144. Springer, New York, 2003.

[75] Elisha Sacks, Leo Joskowicz, Ralf Schultheiss, and Uwe Hinze. Computer-assisted kinematic tolerance analysis of a gear selector mechanism with the configuration space method. In *25th ASME Design Automation Conference*, Las Vegas, ASME Press, 1999.

[76] Elisha Sacks, Leo Joskowicz, Ralf Schultheiss, and Uwe Hinze. Redesign of a spatial gear pair using configuration spaces. In *Proc. of the 2002 Association of Mechanical Engineers Design Automation Conference*, Montreal, 2002.

[77] W. Schiehlen. *Multibody Systems Handbook*. Springer-Verlag, Berlin, 1990.

[78] Joseph E. Shigley, Charles R. Mischke, and Richard G. Budynas. *Mechanical Engineering Design*. 7th ed., McGraw-Hill, New York, 2004.

[79] Lung-Wen Tsai. *Enumeration of Kinematic Structures According to Function*. CRC Press, Boca Raton, FL, 2001.

[80] Herbert Voelcker. A current perspective on tolerancing and metrology. *Manufacturing Review*, 6(4):258–268, 1993.

[81] Daniel Whitney, Olivier Gilbert, and Marek Jastrzebski. Representation of geometric variations using matrix transforms for statistical tolerance analysis. *Research in Engineering Design*, 6(4):191–210, 1994.

[82] Daniel E. Whitney. *Mechanical Assemblies*. Oxford University Press, Oxford, England, 2004.

[83] Kyeonah Yu and Kenneth Goldberg. A complete algorithm for fixture loading. *International Journal of Robotics Research*, 17(11):1214–1224, 1998.

RECEIVED
MAY 12 2010